First Magnitude

A Book of the Bright Sky

First Magnitude

A Book of the Bright Sky

James B Kaler
University of Illinois, Urbana-Champaign, USA

World Scientific

NEW JERSEY · LONDON · SINGAPORE · BEIJING · SHANGHAI · HONG KONG · TAIPEI · CHENNAI

Published by

World Scientific Publishing Co. Pte. Ltd.

5 Toh Tuck Link, Singapore 596224

USA office: 27 Warren Street, Suite 401-402, Hackensack, NJ 07601

UK office: 57 Shelton Street, Covent Garden, London WC2H 9HE

British Library Cataloguing-in-Publication Data
A catalogue record for this book is available from the British Library.

FIRST MAGNITUDE
A Book of the Bright Sky

ISBN 978-981-4417-42-6

Typeset by Stallion Press
Email: enquiries@stallionpress.com

Printed in Singapore by Mainland Press Pte Ltd.

To my family, past, present, and future

CONTENTS

Chapter Five   Bright Star   131

Chapter Six   The Sky is Falling   187

ACKNOWLEDGEMENTS

There are plenty of thanks to go around in the making of this project. First in line is Harry Blom, who directed me to the publisher, and second must be acquisitions editor Jessica Fricchione, who thought this labor of love worthy of publication.

Special thanks go to Mark Killion and Howard Edin for generous use of their images as well as to the Rare Book and Manuscript Library of the University of Illinois for use of a page from the Bayer Atlas. The book would not have been possible without the caring ministrations of senior desk editor Alvin Chong. Final thanks of course go to World Scientific Publishing for their acceptance and fine result. As always I would like to thank my wife Maxine for her help and support.

BRIGHT LIGHTS

When I was young, I lived within a cluster of largish cities. More than half a century ago, lights were dim, skies still dark. From my back yard, I could easily see the Milky Way and stars as faint as my vision would allow. But the lights have had their way, and now seem almost too far gone to stop. They illuminate our towns, and are at times spectacularly pretty as they light big city buildings. But there are always tradeoffs, downsides, to everything. We have light to guide our way many times over, but we've lost the sky which we once could so easily love and admire. I can no longer see the Milky Way — the fabled path of ghostly light that is the combined radiance of the stars of our disk-shaped Galaxy — from my home in a small city. And of course it is in and near the cities where most of us live and work. So the sky appears obscured, leaving us unable to gaze at our awesome Universe.

Or are we unable? There is still plenty to see if only we will look upward, still many things in the sky that are bright enough to cut through the lighted haze, many even visible in daytime, and most important, many that can still awe if we know enough about them. They are all related to things of the "first magnitude," which are the brightest of celestial bodies. We just need to know where they are, how to find them, how to recognize them as they go through their risings and settings and ply their paths across and against the sky and perhaps along the way create other phenomena for us to admire.

"First Magnitude." Everyone knows what it means in popular context: Hollywood's biggest box-office draws, the leaders in the worlds of music, literature, politics. The term comes to us from old astronomy. In the second century BC, Hipparchus of Nicea, among the greatest of ancient Greek astronomers, got the bright idea of putting the stars into six groups, wherein "first magnitude" held the brightest, while those of "second magnitude" belonged to the next rank, and "sixth magnitude" stars were the faintest that a good human eye could see in a dark, dry sky. As a child I could see that far down. Now, with lights and ageing eyes, I am lucky to get to third.

But first magnitude cuts through. The set of objects it contains, visible from almost anywhere (unless you are too far north or south to allow one of them to rise above the horizon), includes 22 stars, some quit colorful, some absolutely amazing in their properties, some duplicitous, others lonely singles, all in their starry way quite lovely. It also includes the five ancient planets, those known since humans first walked the Earth, Mercury through Saturn, one of which can be seen rather easily in broad daylight. Then we add the two brightest bodies of the sky that everyone knows and perhaps loves, the Moon and Sun, the latter by far the brightest of visible stars and so luminous as to be damaging to the eye — which in turn disallows easy admiration — so much so that it turns the dark sky to light, night to day.

These are the permanent residents of first magnitude. There are others that require good luck: bright meteors, "shooting stars," which are pieces of cosmic junk that burst across the heavens through our earthly atmosphere, bright comets, and, among the most awe-inspiring of all celestial sights, exploding stars, which again can be bright enough to be seen in daylight.

In the following seven chapters, we'll look at what these bodies are, how they move, how they appear, how to find and appreciate them, all of them bright enough to keep one looking into even a lighted, citified sky. Most important, we'll examine their significance to us. Stars are more than bright natural lights. Indeed, our very lives depend on them.

The residents of "first magnitude" are all over the sky, as are people on Earth, the apparent sphere of the sky looking very different from the southern hemisphere than it does from the northern. We can't be everywhere, so have to find someplace to stand. For this story, we pick the northern temperate hemisphere, but still travel if need be to see the sights that nature may set out of easy view. *Come along, come look.*

ONE

THE WARMING SUN

I don't know when I first found the Sun, when I first recognized it. I doubt anyone does. It's just always been there for us since memory began. And it will remain so into the immeasurably distant future.

An appropriate chapter number, *ONE*, introduces the one and only Sun. It's hardly the only star, really just one of hundreds of billions in our Galaxy, but it's the one that is ours, that belongs to us, and that gives us light and life. And though we enjoy it immensely, especially on a warm sunny day, on a day in the ball park or at the beach, or rue our sunburn, we pay nearly no attention to it. And with good reason. Unless it is rising or setting through horizon's murk, or is partly obscured by clouds, it's just too bright to look at without damage to the eye. Children, adults too, are warned not to look at a solar eclipse (wherein the Moon cuts in front of the Sun, the phenomenon described in the next chapter), not because of any dangerous "eclipse rays," but because the event draws the eye to the brilliant solar surface, which should never be viewed without a professionally made filter. And even when we can comfortably look at its safely-dimmed disk, we usually see nothing, just a blank yellowish-white circle. Except for a rare giant sunspot, only during a total eclipse when the Sun is entirely covered, allowing the outer solar layers to become visible, do we find anything of interest to the naked eye.

Figure 1.1   Sunrise! The Sun, already blindlingly bright, rises over the distant horizon to begin both the day and our story. We see the Sun through much more air at the bottom than at the top (up to 40 times the thickness overhead). The varying degree of atmospheric absorption gives the gradation in color and brightness. Our blanket of air also refracts (bends) the solar image upward by about its own angular diameter, making it appear higher than it actually is. The upward bending is greater at the bottom than at the top, causing the horizon Sun to take on a squashed appearance. See also Figure 3.8.

What is this remarkable body? Why is it sometimes low in the sky, sometimes high, why does it sometimes rise early, sometimes late? What's it made of, how does it work to warm us? What aspects of it are hidden by its brilliance? In what ways other than heat and light does it affect the Earth? More than we might imagine.

First, What Is It?

A star. But that doesn't tell much. Better to turn it around, as understanding the Sun and its workings is basic to understanding the *other* stars. It's fun to look at its character, its statistics, which are anything but boring. Though appearing deceptively close, it's 93 million miles — 150 million kilometers — away, a fundamental distance called the *Astronomical Unit, AU*. It takes light, at an awesome speed of 186,300 miles (300,000 kilometers) per second, the fastest possible, a bit over eight minutes to make the journey, the Sun, as we say, "eight light-minutes away." (We see the Sun as it was eight minutes ago. Nobody cares.) Even at that distance, it's big enough to appear half a degree across in our sky, its physical diameter measured at 860,000 miles (1.4 million km), 109 times that of our seemingly big Earth. Make the Sun a foot across, and the Earth would be a mote just over tenth of an inch wide a little over 100 feet away.

And it's massive. Squeeze it to that one-foot sphere, put it on an Earthly scale (anything possible in the imagination), and with a mass 333,000 times that of Earth, it would weigh two billion billion billion metric tons. The visible "surface," not solid but a highly opaque gas, is hot: 5500 °Celsius (9945 °F). All that matter is composed very differently from what we have on Earth. Our planet is made largely out of metal — iron and nickel — and rock. Light elements, hydrogen and helium, are scarce. The first is locked up mostly in water, of which there really isn't very much, given that the oceans are but a couple miles deep on a planet 8000 miles (13,000 km) across. The second is far rarer. The Sun on the other hand is made almost entirely of this pair, 90 percent hydrogen, 10 percent helium, and just a salting of everything else, from carbon to uranium.

Most impressive of all is the solar luminosity, the power output. To keep us warm while nearly 100 million miles off, it has to be awesome. And is our star (really a rather modest one) emitting light and heat at a rate of four hundred million billion billion watts? To place the number in a better perspective, let your local electric power company run the Sun for one second, and receive a bill equal to the gross domestic product of the United States for the next million years.

Modern magnitudes place the Sun into context with other heavenly bodies. For more than 2000 years, celestial brightnesses have been measured in Hipparchus's magnitudes (see the Introduction), though now in a form that he probably would have quite enjoyed. The evolving system behaves something like the audio scale of decibels; both more or less adhere to how the senses work, wherein a small change in the number translates into a much bigger change in received power.

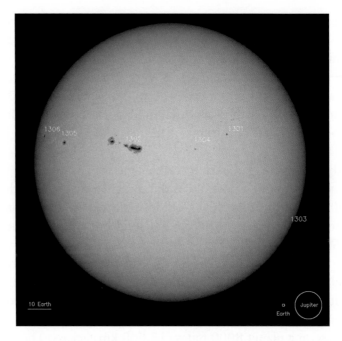

Figure 1.2   The outer Sun. The "surface," the *photosphere*, the "sphere of light," is really an opaque gas made mostly of hydrogen with about a tenth helium. Comparisons to Earth and Jupiter are shown at lower right. Scattered across the solar face are groups of sunspots that are numbered for research purposes. But there is more to see. The apparent graininess of the surface is real, and is caused by upward convection — bubbling — of hot gases. See Figures 1.3 and 1.5. The darkening toward the edge is real as well. The Sun is not a disk but a near-perfect sphere. The outer gases are partially transparent and the temperature increases inward. At the edge of the sphere, you simply do not look as deep as you do at the center, the average temperature along your line of sight is lower, and thus so is the brightness. Solar and Heliospheric Observatory, NASA/ESA.

Modern astronomical magnitudes are adjusted such that five magnitude divisions correspond to a factor of 100 in brightness: a first magnitude star is 100 times brighter than one of sixth magnitude, 2.5 times brighter than one of second magnitude. Magnitudes of course can't just jump from first to second, second to third, etc., but must be continuous. The human eye alone can sense a tenth of a division, and we commonly measure stellar magnitude differences to the hundredth, even to the thousandth. By common agreement, "first magnitude" extends from 0.51 to 1.50, with 1.00 right in the middle, second magnitude from 1.51 to 2.50, and so on.

The relative magnitude scale was originally calibrated by setting Polaris (which is visible any night of the year from anywhere in the northern hemisphere) at magnitude 2.0. The range of the brightest stars is so great, however, that on this scale the top nine stars had to be pushed into magnitude zero (0.50 to −0.49), and even into minus 1 (though generically, these are still broadly called "first magnitude.") Planets can get as bright as minus 5, while the Moon waxes to −12.6. Even at a distance of nearly 100 million miles, the Sun shines on us at magnitude −26.7, which is nearly half a million times brighter than the full Moon and 12.5 billion times the brightness of the brightest star!

The actual solar luminosity is the equivalent of setting off 100 billion one-megaton hydrogen bombs every second, an appropriate analogue, since the Sun in a sense does behave a bit like a hydrogen bomb, but one so tightly controlled that it fortunately cannot ever blow up. And we know why. As a diver goes deeper underwater, the pressure increases as a result of gravity and the resulting weight of the overlying water. An imaginary diver entering the Sun would feel the same. Except that unlike water, the Sun is gaseous instead of liquid, and is compressible, so the density goes up too. At the solar center, this gas — and calculations reveal it must still be one — climbs to a density 160 times that of water, 13 times that of lead! Compress the mixture in a diesel engine and the temperature rises to the ignition point. Compress the Sun using the inward pull of gravity as your piston, and the temperature in the middle also rises, to 16 million degrees C (29 million F). Which, under such high density, is more than enough to ignite the hydrogen, which takes but a few million degrees.

But unlike an earthly engine, there is no combustion with oxygen. It's nuclear fusion, *thermo*nuclear fusion. An atom of hydrogen is made of one tiny particle, a positively charged proton encircled by a neutralizing negative electron. At the temperature of the solar interior, the electrons are whacked off by constant collisions, leaving the protons bare. The higher the temperature of a gas, the faster the atoms move. The density and temperature are so high that the protons can overcome the electric repulsion caused by having similar electric charges. They can therefore blend into a duo (one proton losing its charge to become a neutral neutron) that is held together by a powerful short-range force redundantly called the *strong force*.

Atoms. It's impossible to comprehend how small they are. A proton (or neutron) is but a tenth of a trillionth of a centimeter (1/2.5 inch) across and "weighs" a mere hundredth of a trillionth of a trillionth of a gram. The electron is 0.05 percent lighter, and seems to have no radius at all. Put protons and neutrons together and you create the chemical elements, one proton for hydrogen, six for carbon, 26 for iron, 68 for gold, a rising number of attendant neutrons aiding in the binding of the structures. Wrap them in a number of electrons equal to the number of protons, and you have the various atoms of the chemist's famed periodic table.

This and two more collisions cause four protons to merge together to create helium, which has two protons and two neutrons. The remarkable thing is that, although helium is ultimately made of four protons, it weighs 0.7 percent *less* than the sum of the individuals. We've lost mass! A century ago, Einstein told us where it went in the world's most famous equation: Energy = Mass times the speed of light squared (that is, multiplied by itself) or $E = Mc^2$ (*c* shorthand for light-speed). The missing mass goes flying away as heat and light that slowly work their way to the solar surface, where they are radiated outward. What the Sun is really doing is converting the strong force into supportive and illuminating energy.

We can do that on Earth with hydrogen bombs, in which we start (somewhat differently) upward on the chain, not with simple hydrogen (the necessary heat coming from an interior atomic bomb). It's the first solar step that saves us and makes life possible. It's so slow and hard to achieve that the Sun simply won't explode. These reactions take place only in the deep solar interior, the inner half of the mass and the inner quarter of the radius, where it is hot and dense

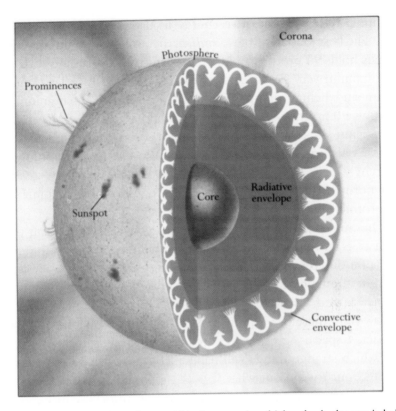

Figure 1.3 The inner Sun. Deep within is a core in which solar hydrogen is being converted into helium. The energy so created supports the Sun and keeps it from collapsing under the weight of its own gravity. The intense heat of the core, which first works its way outward by radiation (through successive absorptions and re-emissions by the atomic gas) and then by tumbling convection, is finally released at the surface in a spray of yellow-white light (see Figures 1.2 and 1.5). Powerful magnetic fields created and concentrated by rotation and convection break through the surface to produce the dark spots and heat a surrounding thin *corona*, which is the seat of the solar wind. From *Stars*, Scientific American Library, New York, Freeman, 1992, copyright J. B. Kaler.

enough. They cannot happen at the surface. All stars work this way, though the reactions might be different.

Heavy chemical elements create energy differently. All the heaviest ones are unstable and, given enough time, spontaneously break apart — *fission* — to lighter elements and dangerous particles with the additional release of radiative energy in the form of ultra-energetic light. The best known of these *radioactive elements* is uranium, whose nucleus holds 92 protons and far more neutrons. Nature fissions ordinary uranium very slowly, and ultimately converts it to lead, making radium and helium along the way. The fissioning of uranium and thorium (90 protons) keeps the interior of the Earth hot, the escaping heat giving us continental drift, volcanoes, and earthquakes; the reactions also making helium for our balloons. Fast fissioning of a different variety of uranium, one with a lesser number of neutrons than normal, makes atomic bombs, as does a form of plutonium (94 protons).

The Sun was given enough interior hydrogen fuel to run pretty much at its current energy output for 10 billion years. The amount of lead (of a different variety than in a fishing sinker, again one with a different number of neutrons) compared with that of uranium in rocks here and elsewhere in the Solar System tells that the age of Earth — and therefore its Sun — is just short of 5 billion years, other radioactive species agreeing. We are halfway through normal solar life, after which strange things will happen. But you'll have to wait until Chapter four to find out what.

And Then, *Where* Is It?

A seemingly silly question. If you cannot find the Sun, you have some serious problems or live under perpetual clouds. Even the profoundly sight-impaired can feel its warmth to find its direction. The question is more about *why* is it seen where it is.

We learn as children that the Sun "rises in the east and sets in the west" in response to the west-to-east daily rotation of the Earth (stars and planets doing the same thing). After it rises, the Sun marches steadily across the sky at a rate of 15 degrees per hour, reaching its peak for the day at noon, when it is at the north-south line of the sky, a circle called the *celestial meridian*. Though it looks like it is the sky rotating around *us*, it's actually *we* who are doing the moving. The motion is just so slow that we do not sense it, and have to look outward to know what is going on.

But now watch more closely, and over the year. Place yourself in the mid-northern hemisphere, almost anywhere in the United States or Europe. On March 20 (the date varying somewhat with leap years), the Sun rises due east, sets due west, is up for 12 hours, down for 12, hence the term *Vernal Equinox* for this first day of spring ("vernal" from Latin for "green"). The point of sunrise then ever so slowly moves into the northeast, sunset into the northwest, until they both reach an extreme on June 20 (the first day of summer, the con-nection to seasons now becoming obvious). During this time, the Sun passes higher and higher to the south at noon, and days are now increasingly longer than nights.

After a brief pause, the pattern reverses itself, until equality again reigns on September 23, the first day of fall, at the *Autumnal Equinox*. Thereafter, we see the opposite, the Sun rising more and more toward the *south*east, setting toward the *south*west, riding the sky lower and lower, days becoming shorter, nights longer, until December 22, the first day of winter. Reversal then follows until we are back at March 20, and the whole process repeats itself. It's some-thing you could watch even from Manhattan.

How come? It's the result of two Earthly motions moving with, or better put, against, each other. The Earth not only rotates once in (by definition) 24 hours, but also — again a childhood lesson — annually orbits the Sun on a closely circular path with a period of a year of 365 1/4 (close enough for now) days. The stars are there for us at night because it's dark enough to see them. Watch them fade at dawn, or emerge at the end of evening twilight, and we know they have to be there in the daytime as well but are just overwhelmed by the brightness of the blue sky.

Sunlight (all light for that matter) consists of nested electric and mag-netic waves moving at, by definition, the speed of light. You see some-thing similar at the beach in ocean water. The color you see in light comes from the length of the separation between wave crests as they hit your eye: red is from longer waves, violet from shorter, and in between we find orange, yellow, green, and blue, and all the other shades. Optical wave-lengths are very small, under a ten-thousandth of a centimeter. The

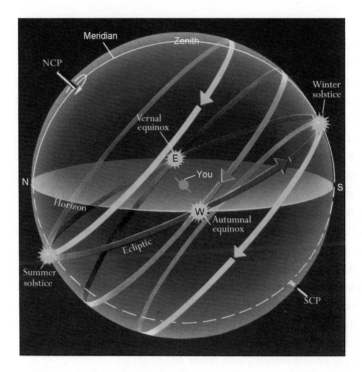

Figure 1.4   Turnings of the sky. The Earth (the small green ball with you "on top") is centered into the sky (big blue ball), which is set for middle northern latitudes in a God's-eye view from the "outside." The Earth's rotational axis points to the sky's two rotational poles, the North and South Celestial Poles, (NCP and SCP), which with the zenith (the overhead point) defines the celestial meridian (the north-south circle). Running parallel to the Earth's equator, the celestial equator (orange circle), arches across the heavens. The compass points are marked on the flat horizon. As the Earth turns counterclockwise on its axis (as looking down from the north), the sky and everything on it appear to turn oppositely, parallel to the equator on the pins of the NCP and SCP (the diagram frozen at an instant in time). While rising and setting daily, as a result of the Earth's orbital motion, the Sun also annually appears to ride the tilted ecliptic (red circle, along which the Zodiacal constellations are strung), in the counterclockwise direction, making it appear to move to the north and south. When crossing the equator at the equinoxes in March and September, the Sun rises exactly east and sets exactly west and seems to move during the day along the equator. When the Sun is at its most northerly position, at the Summer Solstice, it follows the upper yellow circle, rising in the northeast, crossing high to the south, then setting in the northwest. When at the Winter Solstice, however, it moves along the lower yellow path, rising in the southeast, crossing low, and setting in the southwest.

shorter waves carry more energy than the longer ones, violet light twice as energetic as red. Sunlight is a blend of all these colors, which is why it appears more or less white to the eye. Separated by refraction from rain-drops, they create the rainbow (see Figure 1.8). Very energetic light made of extremely short-wave gamma rays is released by fusion within the Sun and by the fissioning of heavy atoms. These solar gamma rays don't make it to the surface, but are degraded by the solar gases into benign sunlight.

The blue sky is caused by sunlight coming at you from all different directions. The air is made of small molecules (combinations of atoms) of nitrogen and oxygen. As they hit the air, some of the Sun's light waves bounce off the molecules. Longer ones run the gantlet far better than the shorter ones: those that your eye sees as violet and blue bounce madly, while those you see as red and orange do not. The eye is not very sensitive to violet, nor does sunlight have all that much violet light to start with (it peaks in yellow). The sky therefore turns out blue from the scattered radiation. And overwhelms the stars.

The blue sky, though, does not just disappear when the Sun goes down. Look into the west after sunset, and there are good odds you can see a jet airplane catching a glint of sunlight that can still be seen from high altitudes. An air molecule "sees" the Sun as well, and can scatter light back to the ground. The sky is not fully dark until the Sun has sunk a full 18 degrees below the horizon, the blue sky gradually fading to black as the Sun descends.

Pretend that you can see the stars in the daytime. They are far away, even the closest of them hundreds of thousands of times more distant than the Sun, most many millions of times. They thus provide a natural and steady backdrop for the Sun. As for rotation, though

Figure 1.4 (*Continued*)   The Moon and planets, which move near the ecliptic, fol-low the same rules. If a star is close enough to the NCP, it will not ever set and is deemed *circumpolar*. If too close to the SCP, however, it will not rise. If you walked north or south on the Earth, the sky would appear to shift oppositely. From the southern hemisphere, it is the SCP that is up, the NCP down. Adapted from *Stars*, Scientific American Library, New York, Freeman, 1992, copyright J. B. Kaler.

the Earth orbits at a speed of 18.5 miles (29.9 km) per second, the motion is so smooth that we don't feel it. So instead, it looks as if the Sun is annually going around the Earth: a concept believed until Copernicus took it away in the 16th century and told us that the Earth, and all the other planets, go around the Sun instead.

If we could watch, we'd see the Sun move along a fixed path against the distant stars, a path with a special name, the *ecliptic.* Since the Earth orbits counterclockwise as viewed from the north, the Sun seems to move annually in the same direction, to the east, against the starry background (while at the same time moving daily clockwise, from east to west). The stars are scattered across the sky more or less at random, and naturally make patterns that the ancients named and made stories about, and that we call *constellations.* Since the Sun-god moved in the sky, those that held the ecliptic must be special and sacred. Indeed, these dozen constellations of the *Zodiac* are all related to living things, the very name from Greek meaning "circle of animals." (Libra is the exception. Scales are not alive. It was once the outstretched claws of the Scorpius, the Scorpion next door to the east, maintaining the concept of the living circle.) To the ancient mind, the other moving bodies of the sky, the Moon and planets, also represented gods. All are also found against the zodiacal constellations. Their relations to the constellations (and the characters they represent) and to one another might be then be used to tell fates and fortunes, hence the rise of astrology, the "signs" of the Zodiac originally taken from the astronomical constellations.

The sky's 48 ancient constellations (see Chapter 4) include such beloved figures as Ursa Major (the Greater Bear), which holds the Big Dipper, and Orion (the Hunter) with his prominent three-star Belt. Official adoption of 38 "modern" patterns and the breakup of Argo (of Argonaut fame) then brought the total count to 88. The special constellations of the astronomical Zodiac (as opposed to the astrological signs), and the average dates of solar residence within their modern boundaries, follow in traditional order, from west to east. There are twelve because to the nearest whole number, the Moon the goes through its phases 12 times per year, each phase more or less successively taking place within the next one to the west.

Aries	Ram	April 19–May 14
Taurus	Bull (Fig. 3.2)	May 15–June 20
Gemini	Twins (Fig. 4.2)	June 21–July 20
Cancer	Crab	July 21–August 10
Leo	Lion (Fig. 5.15)	August 11–September 16
Virgo	Maiden (Fig. 5.1)	September 17–October 31
Libra	Scales	November 1–November 23
Scorpius*	Scorpion (Fig. 5.2)	November 24–December 17*
Sagittarius	Archer	December 18–January 19
Capricornus	Water Goat	January 20–February 16
Aquarius	Water Bearer	February 17–March 11
Pisces	Fishes	March 12–April 18

The boundaries of Ophiuchus (the Serpent Bearer) cut across the ecliptic. This so-called "thirteenth constellation of the Zodiac" contains the Sun from November 30 to December 17; these passage times are combined with Scorpius (which sees the Sun but for five days). Nevertheless, Ophiuchus is not part of the classical Zodiac.

The Earth rotates around an axis that emerges at the north and south poles, which respectively lie in the Arctic Ocean and in the high plateau of central Antarctica. Exactly in between them lies the circle of the equator. The sky also has north and south *celestial* poles and an equator that lie above their respective earthly counterparts. As the Earth turns on its axis, the sky appears to rotate around the celestial poles parallel to the *celestial* equator. The Earth's orbit also has an axis, the perpendicular to the orbital plane at the Sun. If the rotational and orbital axes were parallel, that is, if the Earth stood "upright" relative to its orbit, the ecliptic would lie on the celestial equator, the Sun would shine perpetually over the Earth's equator, would always rise exactly east, set exactly west, and days and night would always be equal.

But, and here is the perpetual sticking point, it doesn't. The two axes are *not* parallel. Instead, the rotational axis is, by accident of nature, tilted over relative to the orbital axis by 23 1/2 degrees (more precisely, 23.4). This tilt causes the ecliptic to be inclined to the celestial equator by the same angle. As the Sun rides along its ecliptic path, it is then, while moving easterly, also forced to go first toward the north and then to the south, crossing the equator twice a year at the equinoxes, which are actually points in the sky. When it is at its most northerly position, at the *Summer Solstice*, 23.4 degrees north of the equator, the Sun rises in the northeast, sets in the northwest, and crosses to the south very high in the sky. When at its most southerly position, 23.4 degrees south of the equator, it rises and sets in the southeast and southwest and crosses the sky low. It's all diagrammed in Figure 1.4. The Summer Solstice is in the direction of the constellation Gemini (the Twins), while the winter one is "in" Sagittarius, the Archer. Vernal and Autumnal Equinoxes are in Pisces (Fishes) and Virgo (Maiden).

Other Places

Now travel. Another big city is fine. From Earth's mid-southern hemisphere, all is reversed. Our Summer Solstice is their winter one, and so on. To see how that happens, go back to your station in the north. The height of the Sun at noon depends on where you are. Stop the Earth's rotation right at that moment. Walk south along the spherical Earth and the Sun moves north. Terrestrial latitude is the positional angle north or south of the equator, that of the equator 0°, of the north pole 90°N, south pole 90°S, of New York 40°N. When in your southbound travel you hit the magic latitude of 23.4 degrees north, the Summer Solstice passes overhead. And on the first day of summer, June 21, so does the Sun! You are now at the *Tropic of Cancer*. If you are at the equator, then the equinoxes pass overhead, as does the Sun on March 20 and September 23. Here, and only here, are the days and nights always of equal length. At 23.4 degrees south latitude, you can look up to admire the Winter (as those in the north would call it) Solstice passing overhead along with the Sun on December 22. The *tropics* (referring in the original Greek to the "turning" Sun as it reverses its north-south direction) are where the Sun can pass overhead. You will never see it that way from New York or any other place in the United States except Hawaii. (Canadians are out of luck too.)

Why Tropics of Cancer and Capricorn, when the Summer and Winter Solstices are in Gemini and Sagittarius, each one zodiacal constellation to the west? The Earth has a THIRD motion, a wobble in its rotational axis, called "precession." It's the same phenomenon as the wobble of a spinning top. The tilt of the rotation axis stays about the same, but the axis changes its direction in a circle around the orbital axis over a long period of 26,000 years. The axis toward north now roughly points to the second-magnitude star Polaris (the fabled North Star), but that is only temporary. As the axis wobbles, the equinoxes and solstices shift as well, to the west against the stars at an average rate of about one constellation of the Zodiac every 2000 years or so. When our western constellations were being invented, the Summer Solstice was indeed in Cancer, the winter one in Capricornus. The astrological signs are, however, strangely linked to the equinoxes and solstices, by now have each shifted one astronomical constellation to the west. Hence the names of the Tropics. In parallel, the Vernal Equinox was once in Aries, which is why it usually tops the newspaper horoscope lists, while the autumnal equinox was in Libra, reflecting the "balance" of days. Precession was discovered with the naked eye by none other than Hipparchos of Nicea, he who created the magnitude system that led to the title of this book.

The equator is a special place. At other mid-northern latitudes, the Sun does not set directly down, but at an angle. The closer you are to the equator, the more vertically it descends. At the equator, the solar daily path is perpendicular to the horizon, the only place on Earth where that happens (the reason for the equality of days and nights). The duration of twilight depends on how long it takes the Sun to reach the critical 18 degree limit below the horizon. The tropics therefore have shorter twilights than does New York, while at the equator, they are the shortest of all, which can quite fool vacationers out for an after-dinner walk.

Now go home to your mid-northern latitude and walk *north* on June 21. The points of sunrise and sunset creep ever more northeasterly and northwesterly as the solstice noonday Sun sinks continuously farther down. The angle of the sunset path lowers and twilights lengthen. Eventually, at a latitude of 66 1/2°N (90 minus 23 1/2), the rising and setting points meet at due north and the Sun stays up for 24 hours. From this place, from the *Arctic Circle* and up, you can witness the midnight Sun on June 21. The farther north you go, the higher the Sun at midnight and the more days there are of continuous sunlight. At the north pole, we see the Sun up for six months straight, from March 20 to September 23.

But gain some, lose some. North of the Arctic Circle, the Sun around wintertime does not rise, and at the pole we witness six months of gloom that consists of a couple months of deepening twilight followed by a couple months of true night that are in turn followed by a another pair with brightening twilight. Below the Antarctic Circle (66 ½°S) to the South Pole, all is reversed. There is nothing hidden here; no telescope is needed. All these phenomena are readily visible to the naked eye.

You don't really have to go as far as the Arctic Circle to witness the midnight Sun effect. If the Sun gets less than 18 degrees below the horizon at midnight, you will see some twilight. As far south then as 49 or so degrees north latitude, there will be a glow in the north all night long, one that gets stronger the farther north you go. The effect covers most of Canada and northern Europe. Hence St. Petersburg's famed "white nights."

The Sun not only emits the colors of the rainbow, but low energy long-wave "infrared" radiation you feel as heat, and high energy "ultraviolet" light that has wavelengths just shorter than violet (but far longer than gamma rays and the X-rays that lie between them). Ultraviolet is dangerous and can burn the skin. The worst of it is absorbed by the Earth's atmosphere (mostly by high-altitude ozone), but a small bit sneaks through. The fastest route to space lies directly overhead. As you look toward the horizon from the overhead point, you look through more and more air until at the horizon the atmosphere becomes almost

40 times as thick (see Figure 1.1). When the Sun is near the horizon, the absorption of ultraviolet light is so fierce that none gets through, while the overhead Sun beams down the maximum. Where then so you get the worst sunburns? In the tropics, which is sometimes a stunning, and oft-times painful if not dangerous, surprise for vacationers.

Seasonings

The Earth's orbit is not quite circular, but a bit elliptical (as are the orbits of all the other planets) with the Sun offset some from the center (if you know your ellipses, at a focus). With close attention, you can actually "see" the non-circular effect, and in two quite different ways. First way: we are 1.7 percent (1.6 million miles, 2.5 million km) closer to the Sun around January 2 at a place called *perihelion* (after Helios, Greek god of the Sun), farthest (at *aphelion*) around July 3. The Earth's orbit is controlled by the gravitational attraction between it and the Sun. If you get closer, gravity goes up, and the Earth moves faster (a discovery made by Johannes Kepler in 1610). Since perihelion and aphelion passages are accidentally close to the times of the solstice passages, the Sun hurries through northern winter and fall, lingers in spring and summer, these latter seasons some three days longer than the other two. It may not seem like it, but count the days in the calendar. The ancient Greeks actually noted the effect, but had no idea of why it was so.

But something appears wrong. The first thing we might think of as the cause of seasons is solar distance. But the Sun is closest to us in the dead of northern winter. Moreover, when it is summer in the northern hemisphere, it is winter in the southern, and vice versa. So that explanation seems not to work. Why then the seasons? The cause is the ever-present tilt. The heating rate of the ground beneath you depends on the angle with which sunlight hits it. Maximum heating would be with the Sun directly overhead. Take the Sun from the overhead point, and the solar rays must spread themselves out over more area of ground, and the heating rate per square mile, kilometer, inch, whatever unit you choose, goes down.

The only thing that really counts is how high the Sun is on the meridian at noon; the lower it is, the cooler it will be outdoors. At the north and south poles, the Sun can be up for 24 hours straight, but can never get more than 23 1/2 degrees above the horizon. So it stays cold. In moderate climes, the Sun is higher at noon in summer than in winter, so it's hotter; that is why we *have* summer and winter. The hottest places on Earth are in the tropics, where the Sun is most nearly overhead.

We are closer to the Sun in southern hemisphere summer than in northern hemisphere summer, so all things equal, the former is hotter than the latter. But on Earth, all things are *not* equal. There is far more heat-moderating water in the southern hemisphere, and the effect is thereby lost.

Time

The second easy way of "seeing" orbital eccentricity involves clocks. The time of day is rigged to the position of the Sun. Our clocks represent the Sun's movement across the sky. When the Sun hits its highest peak for the day, our time is "12 noon." For every 15° (360/24) of westerly solar motion across the sky, we add another hour to the clock. After turning 90°, it's six hours later, 6 PM ("post meridiem," Latin for "after noon"), or on the 24-hour system, 18 hours. After 12 hours it's 12 midnight, and we go back to zero. After 270° it's 06 hours, or 6 AM (ante, before, meridiem). This is the time you would (in the daytime) read on a sundial properly subdivided by minutes. A source of considerable confusion is that 12 noon and midnight by definition have no AM or PM.

But then there is the elliptical orbit to contend with. Taking 365 1/4 days to go around the ecliptic, the Sun must move against the stellar background at just a hair under 1 degree (0.986°) per day. The reason there are 360° in a circle is because the Sun takes 365 or so days to make its apparent circuit, and 360 is the closest, most easily manipulatable, number. But in January, the Earth moves faster than normal, so the Sun appears to move more quickly east along its apparent path. The length of day is from noon to noon. But if the Sun is moving faster to the east along the ecliptic and counter to its apparent movement to the west across the sky, the full day is longer than average, while in July, with the Sun moving slower, the day must be shorter. But we still divide the day into 24 hours, so the simple solar hour is longer in northern winter than it is in summer! The effect is complicated by the Sun moving along a path that is inclined to the equator.

While quite possible, it takes some fancy clockwork to keep up with raw "solar time." Even if done, in a modern world we can't have a variable time counter. We need a steady tick (these days, hum) of the clock. So we average out the vagaries of movement, and calculate a "mean solar time" on the basis of a fake mathematical Sun that moves steadily along the sky's *equator*.

As the year wears on, the real Sun, the true yellowish orb, periodically gets ahead of, then behind, its average projected equatorial position. An extreme occurs late in the year, when the discordance hits around a quarter of an hour, with the real Sun well to the west of the steady-paced one. The result is that while the shortest day of the northern hemisphere year falls on December 22, the earliest sunset as told by the mean solar clock takes place on December 7, more than two weeks before Winter Solstice passage, by which time the sky is notably lighter in the late afternoon. Symmetrically, the latest sunrise occurs a couple weeks after the solstice. Both effects, this and the related "inequality of the seasons," are quite easily sensed with no instrumentation at all — from anywhere.

Days and Years

Earth rotation takes 24 hours relative to the Sun. The day and the year, however, are entirely independent of each other, meaning that there cannot be an exact number of days per year. Which massively complicates calendars, which are designed to keep track of days and fit them properly to within the year. The actual orbital period of the Earth around the Sun is 365.24219... days, the little dots meaning that the number has a continuous run of decimal places (to the mathematician an "irrational number"), a hair under the usually-expressed 365 1/4.

Different cultures solve the problem in different ways, some of them involving the additional complication of the Moon (Chapter 2).

Our strictly solar calendar owes a large debt to Julius Caesar's astronomer, the Greek Sosigenes. In reforming the ancient Roman calendar, he simply adopted the *leap year*. For three years running, the year has 365 days. In the fourth year, an extra day is added (as our February 29), so that the average over four years comes out to the standard 365.25 days. For Roman reasons, he also placed the date of the Vernal Equinox passage on March 25.

As good as it is, however, the *Julian calendar* is too long by 0.0078 days, or nearly 12 minutes, per year, which causes the time of the Vernal Equinox passage to move forward in the calendar by one day every 128 years. By the late 16th century, the first day of spring was falling on March 12, which was seriously out of line with religious values and views. So when Pope Gregory XIII ordered a calendrical review and revision, his two astronomers suggested two measures. First, drop leap years in century years not divisible by 400. The years 1800 and 1900, though falling into the four year cycle, were not leap years, while 2000 *was* (to the consternation of some calendar makers who got it wrong, thus showing that you need to take your astronomy seriously). Removing three leap years out of 400 years gives a long-term average of 365.2425 days, so close that hardly anybody cares about the difference.

Gregory's astronomers also reset the first day of spring to March 21, where it was at the time of the religious Council of Nicea in the year 325, thus dropping more than a week from the calendar. By the time that England and its colonies figured out that the new calendar was a good idea in spite of coming from Rome, it was 1752, and the Julian system was even more out of synchrony with the seasons. The move from the Julian to the *Gregorian calendar*, from "old style" to "new style" as it was once called, resulted in riots by people who thought days were being stolen from their lives. The next chance you will get in the exercise is the year 2100, which again will skip the leap year.

The day is just as complicated. Our *solar* day runs from noon to noon, when the Sun reaches its highest peak. The *mean* solar day simply uses the fictitious mean Sun that rides the equator. But we need contend with the Earth going around the Sun, which causes our visible target to move to the east at a rate of about a degree per day. Imagine a star and the noonday Sun along the same line of sight. Both move with the Earth's rotation to the west toward the setting point. But at the same time, the Sun moves to the east of the star. The star then returns to its highest arc *before* the Sun to define a *sidereal* (from the Greek meaning "star") *day*. At that moment, the Sun is still a degree east of its noon position, and it takes another four minutes to return. The sidereal day, four minutes (actually 3 minutes 56 seconds) shorter than the solar, is the true rotation period of the Earth, one that a Martian would measure.

Then divide both the solar and sidereal days into hours, minutes, and seconds. The sidereal day must cram 24 hours into a shorter interval, and both they and the sidereal minute and second must be shorter as well. The start of the civil day, zero hours, is midnight, while that of the sidereal day is the moment the Vernal Equinox crosses the meridian to the south. The two clocks read the same at the instant the Sun crosses the autumnal equinox. The sidereal then gains not quite four minutes per day over the solar. All observatories maintain both kinds of time on solar and sidereal clocks (even if they are buried as software in the telescope operating systems) in order to find things in the sky. Because it is easier to measure the positions of stars in the sky than it is the Sun, *sidereal time* is the actual world measure, not solar, the former converted into the latter to tell our civil time of day.

The effect is easily seen at night. Pick a time. Look to the south. At midnight you see stars that are at the moment opposite the Sun. As the Earth orbits and the Sun plies its apparent ecliptic path, each night you face in a different direction, by one degree farther to the east. As the year progresses, the midnight sky (or what you see at any specific time at night) slowly shifts to the west. The sky then appears to do a full extra rotation over the course of the year, the winter constellations being replaced by those of spring, and so on, giving us 366 1/4 sidereal days per year. In the old days of precision pendulum clocks, you could listen to the ticks of the sidereal clock catching up to, and then passing, those of the solar. You could *hear* the Earth going around the Sun. And be indoors too.

Action!

When the Sun is low, and the horizon is thick with moisture, dust, or exhaust fumes that sufficiently dim the solar disk, you can look directly at it, and once in a great while you can see a naked-eye sunspot. Galileo is said to have discovered sunspots with his small telescope, but the record goes way back to ancient China. Through the properly filtered telescope, sunspots — dark markings on the Sun — can be spectacular. They are said in our lore to influence plant growth, climate patterns, even (stretching credulity) the stock market. As we watch, the spots come and go, some lasting for mere days, others for up to a couple months. The numbers of spots are also cyclic over a much longer period that averages 11 years. At the peak of the *sunspot cycle*, there may be dozens of them, while at minimum we might go for a couple years during which we see hardly any at all.

Sunspots are cooled, more or less circular solar regions that range in size from barely-visible motes to huge blemishes considerably bigger than Earth. The amount of light emitted by a body depends critically on its temperature. Though the spots are only about 1500 °C cooler than the surrounding gaseous "surface," they appear black by contrast. Contrary to their reputations, they actually do nothing at all. Collectively, they are just one more symptom of solar magnetism, the real source of solar activity, and one not directly sensed by eye but only indirectly by its effects.

Like the Earth, the Sun rotates, though much more slowly, taking nearly a month to make a full turn as witnessed by the seeming march of sunspots across its surface. (Given a naked-eye spot and an appropriate filter, you could watch it yourself.) But unlike Earth, it does not rotate solidly. The equator takes 25 days, while near the poles it takes closer to 30, the solar gases shearing past each other. (If the Earth did this, Central America would at times wind up south of Europe, Kansas south of Maine.) Precision telescopes also show the solar surface to be broken into millions of bright *granules* that rise, radiate away their heat, and fall, as the Sun seems to boil like a pot of water in a process called *convection*. Not so much a surface phenomenon as it is global, this mad circulation, which moves the gases at several

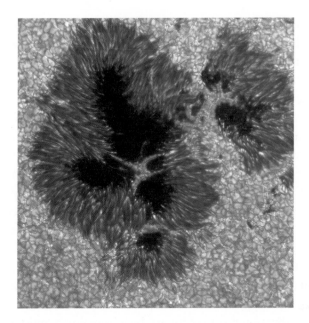

Figure 1.5   Grains and spots. The photosphere is broken into a million or more tiny granules, "rice grains," that are the tops of roiling convection cells each about the size of a large US state. Rising quickly, a granule lasts only about 10 minutes before radiating its heat, the cooled gases then dropping down along the grain's old dark border. Set into the scene is a closeup of a large sunspot group. Each spot is made of a dark, cooler, depressed *umbra* that is surrounded by a lighter, striated *penumbra* that slopes back upward to meet the granulated "surface." Created by intense magnetic fields that loop out of the Sun and inhibit convection, spots are paired like bar magnets, one positive, the other negative. Even big spots like these, which could easily swallow Earth, last no more than a month or so. Mats Löfdahl with the Swedish 1-m solar telescope, Institute for Solar Physics, Royal Swedish Academy of Sciences.

hundred meters per second, extends a third of the way into the Sun, the gases below the convection zone remaining quiet until we reach the nuclear-burning core (see Figure 1.3).

Since inside the Sun, electrons can be stripped from atoms as a result of elevated temperatures, the solar gases are electrified. Movement of a charged particle creates magnetism. Combination of all these motions thus produces a global solar magnetic field vaguely similar to that of Earth's, but one that the non-solid rotation ties up in thick ropes that bubble to the surface. Where they emerge in gigantic loops, they quiet the convection and refrigerate the surface to create a pair of spots. The magnetic fields are ephemeral, and therefore so are the spots.

There are other aspects of the Sun that tie into the spots that are not readily visible to the eye, but whose effects affect us in profound ways that can be easily sensed. As the Moon goes around the Earth, it occasionally passes directly in front of the Sun. By a wonderful coincidence of nature, the Moon and Sun have the same angular diameter of 1/2 degree, so if conditions are right (see Chapter 2) the lunar disk can exactly blot out the solar. When we fully eclipse the brilliant solar disk, what emerges is a huge, pearly, circumsolar halo, the *solar corona* (see Figure 2.10). Various measures show it to have a temperature of more than a million degrees, but it's so thin and vacuous that it radiates hardly at all and, like the stars, cannot be seen in a bright blue sky.

Though the exact source of its heat remains a mystery, we do know that the corona is caused by the release of solar magnetic energy. Magnetic activity also generates harsh ultraviolet radiation, even X-rays, that can be dangerous to space travelers. Like any hot gas, the corona tries to expand, and quickly. It can be held back by the same magnetic loops that create the spots, but where it is not (and to a lesser degree even where it is), the corona flows rapidly outward and becomes the *solar wind*, which blows past us at up to hundreds of miles (and kilometers) per second and pervades the whole Solar System. The Sun is evaporating, though we need hardly worry, as at that rate it would take 100,000 billion years to dry up completely, and the core will die long before that (giving us something else to worry about instead).

The energy of the raw solar wind is dangerous, and is one of the many hazards of space travel. We are protected from it on the ground by our atmosphere, and particularly by our magnetic field, which is generated by the circulation of Earth's deep, liquid iron core. Our magnetism traps solar wind particles in a pair of donut-shaped rings, the *van Allen radiation belts*, one of which is 1.5 Earth radii out, the other four. As the solar wind moves outward, it drags along with it

Figure 1.6   Aurora. The northern lights, a near-perpetual feature of the far north, shimmer at lower latitudes thanks to intense solar magnetic activity as seen in the next illustration. The constellation Gemini (see Figure 4.2), with its bright stars Castor (on top) and Pollux, lies at center right, with Saturn falling below them.

the Sun's magnetic field, which can connect to the Earth's field. The whole affair generates wide rings of electrical current around the magnetic poles (where our field emerges) and that are offset from the rotation poles by some 10 degrees (the northern pole in far northern Canada, which is where a magnetic compass will take you). The current ring causes the upper atmosphere to glow quite vigorously, and creates the *aurora*, or northern lights, which are readily visible in the lands beneath the ring, including Alaska and northern Canada (the southern lights glowing symmetrically in and around Antarctica).

Back now to the Sun. When the fragile magnetic loops cross over and neutralize one another, they collapse. The released energy can send particles whipping down to the solar surface beneath them to create bright, localized *solar flares*. More, the part of the corona they have been confining explosively expands. A blob of hot gas, a *coronal mass ejection*, then comes hurtling down the solar wind and, if we are in the way, we get smacked by it. The event (and very high-speed particles related to flares on the solar surface) massively disrupts the radiation belts and current rings, causing them to expand, and suddenly the aurora is visible in lower temperate latitudes, on occasion well into the southern US. The pulsing glow, colored with reds, greens, blues, can sometimes be seen from towns so bright that even the stars might be heavily obscured.

There are also hidden effects that are sensible even in the biggest of cities. The additional electrification of the Earth's upper atmosphere can seriously damage radio communication. Moreover, electricity and magnetism are flip sides of each other. The electric currents that make the aurora generate magnetic fields that stretch to the ground. They in turn generate weak ground currents whose own magnetic fields can disrupt long-distance power lines. Circuit breakers blow, and your city — perhaps your whole state or province — suddenly goes dark. It's happened. And your taxes go up to pay for repairs. And you buy fewer stocks. And all because of sunspots — rather the amazing, near-hidden phenomena that make them. On the other hand, when the lights do go out, be sure to admire all the stars and even the Milky Way! There is a story, perhaps apocryphal, that when a power failure made Los Angeles go dark, people got scared of all the little lights in the sky and called the police. Oh what we have lost.

There are other economic effects that affect our modern world. A coronal mass ejection can knock out both commercial and military satellites. Moreover, friction from the residual high upper atmosphere slowly degrades the orbits of near-Earth satellites, eventually bringing them down. Solar activity heats the atmosphere and makes it expand, resulting in increased degradation, and there goes another billion dollar orbiter. And your taxes go up even more.

The aurora and all the related phenomena follow the 11-year solar cycle. Between roughly 1645 and 1715, the cycle effectively shut down, during which the northern world was plunged into a period of intensely cold weather. The event, called the *Maunder minimum* after its discoverer and its purported result, may be coincidental, but data from tree rings and ice cores, which respond in various ways to the solar cycle, show that it has happened before. The Sun and the solar cycle thereby enter strongly into the politics of global warming.

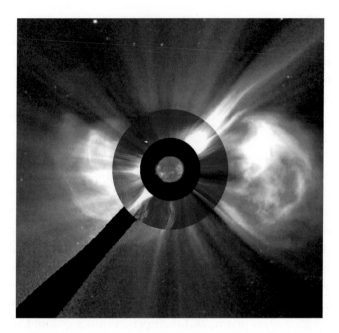

Figure 1.7 Explosion. An X-ray view of intense magnetic loops in a hot layer just above the solar photosphere (the small circle) is set into the middle of the composite image. Surrounding the blank circle, two more instruments take over to show (as the eye might see it) the huge two-million-degree solar corona, which is normally confined in much larger magnetic loops. When the magnetic fields neutralize each other and collapse, giant coronal bubbles explosively take off as they do here to the right and left. If the Earth is in the way, they disrupt our own magnetic field and produce strong auroral displays that can extend to low latitudes (as in Figure 1.6). Solar and Heliospheric Observatory, from the ESA/NASA Helioviewer Project, D. Müller (ESA), J. Ireland (ADNET Systems, Inc/ NASA Goddard Space Flight Center), and an interdisciplinary team.

Under the Rainbow

Air molecules scatter blue and violet light out of sunlight to make the blue sky. Blue light is also absorbed better than red. The Sun therefore must look a bit redder than it would from space. The effect when the Sun is high is to help give it a slightly yellowish cast. But as the Sun sets, sunlight has to pass through more and more air, so the Sun grows more and more golden. At the horizon, where the pathway through the atmosphere is maximized, the Sun sometimes turns a deep orange or even red (see Figure 3.8). The effect in twilight is exaggerated even more by a Sun *below* the horizon. The result may be a beautifully colored sunset (or rise) that can be stunningly enhanced by reflection of sunlight from clouds.

While waiting for sunset, a cleansing afternoon thunderstorm flees off to the east. Suddenly, against the falling rain is the traditional message of hope, a semi-circular ring of colored light, a rainbow. It's the best-known of a huge set of interactions between sunlight and air other than the blue sky itself. Among the most common of optical phenomena is *refraction*, a bending in the direction of light waves as they pass from one medium or substance (or a vacuum) into another. Fish look funny and distorted while swimming in their tank as the light passes from water, through glass, and into the air. And you look just as funny to the fish. The degree of bending depends on the light's wavelength, hence color, blue and violet more affected than red. Pass light through a glass prism, and it disperses into a colored spectrum. Natural prisms are everywhere. Just look at the colored refractions of sunlight on fresh snow.

Water droplets make another fine example. Sunlight falling on a raindrop refracts and disperses into its component colors upon entry, bounces off the droplet's backside, then refracts and disperses again upon exit. The sum of all the refractions and reflections from countless drops then creates a circular colored ring around the point opposite the Sun 42 degrees in radius with red on the outside, blue, even violet, on the inside. A double reflection inside the drop makes a fainter outer ring 51 degrees in radius with the colors reversed.

Ice works well too. Light cirrus clouds form high, tens of thousands of feet up, where the air is cold and water can freeze into tiny six-sided crystals. Like the leaded glass drops of an expensive chandelier, they refract and disperse light, in the natural world, sunlight. The result can be a magnificent colored ring 22 degrees in radius, red on the inside, blue outside, with the Sun smack in the middle. (Be sure to hide the brilliant Sun behind a tree or building.) When the Sun is low, the portions of the ring on a line through the Sun parallel to the ground can get very bright. Setting (or rising) along

Figure 1.8   The beauty of the rainbow tells that the white light of the Sun consists of myriad shades of red, orange, yellow, green, blue, and violet. The brighter inner bow, whose center is opposite the Sun, is caused by sunlight that refracts at entry into raindrops, bounces off their backsides, and refracts again into the observer's eye. Double reflections inside the drops make the fainter outer bow.

with the Sun, these *sundogs* (*mock suns, parhelia*) can be seen even without the ring, and can be so bright to be uncomfortable to admire them.

Smaller colored arcs can appear atop or alongside the big halo. On rare occasions the crystals can produce an outer ring 47 degrees in radius along with its attendant tangents that include a partial ring, not often noticed, about the overhead point. A white ring might also pass through the Sun around the whole sky parallel to the horizon, the heavens becoming a great colored canvas under Nature's brush.

More common, yet no less striking, are cloud shadows thrown into dusty or humid air. From the observer's perspective, the parallel sunbeams not blocked by the clouds appear to diverge outward to make a glorious "sunburst" effect. Even more common is atmospheric refraction acting on the rising and setting Sun, lofting it upward by a full half degree and squashing the apparent solar disk, as vividly seen in Figures 1.1 and 3.8. Indeed it's a daily affair, extending the duration of daylight.

And you can see it all from just about anywhere. Even under Moonlight, to which we now turn.

TWO

THE CHANGING MOON

As for the Sun, there are few memories of discovering the Moon. My most durable recollection is of seeing it through speeding clouds. Children love it, my oldest daughter, then age two, saying "mun" "mun" over and over when she saw it. There is little more magic than waking in the early hours and seeing the full Moon flooding the landscape with a silvery light. Composers and poets would be lost without it.

The chapter number is again apt, the Moon the number two body of the sky, one that might be seen and admired at any time of the day or night from anywhere, country to Chicago. Unlike the Sun, it constantly changes its apparent shape. While sometimes nearly invisible, it can be so bright that it allows us to find our way in the dark and almost to read. At its best, the Moon hits magnitude −12.6, not quite a half-millionth as bright as the Sun.

While still nearly 30,000 times brighter than the brightest star, the Moon is not the least bit dangerous to look at. Shining only by reflected sunlight, the Moon is really only as bright as dark earthly rock seen under a clear daytime sky.

The Moon is not only artistically inspiring, but causes profound effects on Earth, including the perpetual tides, and seems to play a role in life's very existence. On the other hand, its lovely light can certainly annoy astronomers seeking dark skies. As with the Sun, the Moon is related to the gods. Reflecting our cultures (as well as

Figure 2.1   A waxing crescent Moon glides down a western twilight sky. Positioned some 30 degrees to the east of the much-more distant Sun, the Moon shows us a sliver of her daylight side, with her nighttime side aglow with sunlight reflected from the bright Earth. Measures of earthlight are used to study the global reflectivity of our planet, including cloud cover.

sunlight), the Sun is male: Apollo or Helios depending on the times and stories; the Moon female: Artemis (Greek) or Diana (Latin), she and Apollo twins.

The Moon, *the* Moon rather than just "a moon," is our one and only natural satellite, a body that is more beholden to us than to the planetary system at large. It is to Earth as Earth is to Sun, taking a month to make an orbit around us. Two thousand two hundred miles (3476 km) across, averaging 239,000 miles (384,400 km) miles away, it's just over a quarter the diameter of Earth, making it half a degree wide in our sky, by a rather remarkable coincidence the same as the Sun.

Gravity and orbits. Isaac Newton found that two masses of any sort will attract each other with a force proportional to the product of their masses divided by the square of the distance between them (rather, between their centers). Double the distance, and the gravitational strength decreases by a factor of four. The Earth and Moon then mutually attract each other, and thus try to fall together. The Moon, however, is moving in a direction more or less perpendicular to the direction to Earth. Though the Moon is trying to fall, the horizontal motion is so fast that the Earth curves beneath it, and the Moon cannot catch up to us.

All orbits work this way, including those of artificial satellites. To launch one, you fire it high enough so that the residual atmosphere is not much of a problem, while at the same time accelerating the rocket to a speed of 18,000 miles (29,000 km) per hour parallel to the Earth's surface. Cut and eject the rocket motor and release the satellite. The artificial Moon then falls in the Earth's gravity, but it's going so fast it can't descend. Astronauts feel weightless because they and their spacecraft are falling together in the same way. To bring down the spacecraft, fire a rocket in the forward direction to slow it down, and it spirals back to Earth.

Moons are common in the outer Solar System, Jupiter and Saturn having dozens. Among the inner planets, however, it's the only one of any size (Mars has a pair of tiny ones), its mass just over one percent of Earth's. The Moon is big enough that those who study it see it not so much as a satellite but as our planetary companion, making the Earth–Moon something of a "double planet." Behaving very much like a real planet, the Moon is in many ways similar to Earth; on the other hand its smaller size has led to some profound differences. As here, you could easily walk the rocky surface, but you'd need a spacesuit and an air tank to keep you alive. Not to mention a bottle of water.

Earth First

We cannot understand the Moon without first looking at Earth. It provides the paradigm, the prime example, not just for the Moon, but for the other three inner planets as well. It's basically a ball made of rock and iron 7900 miles (12,750 km) across. In the inner half of the diameter is a *core* that consists mostly of a mixture of 90 percent iron and 10 percent nickel that constitutes a third of the terrestrial mass of 6000 billion billion metric tons. The inner half of the core is solid, while the outer half is liquid — a consequence of the balance between heat and pressure (the central temperature a whopping 7000°C, hotter than the surface of the Sun).

Surrounding the core is a thick rocky coat — the *mantle* — that constitutes almost all the rest of the radius and mass. It's made mostly of various metal-holding silicates and carbonates into which are mixed nearly all of the chemical elements. Pretty much missing, however, in both core and mantle, are the really light ones like hydrogen and helium, which in a global context are preciously rare. Most of the hydrogen is locked in water, and all of the helium is a by product of the radioactive decay of uranium and thorium — which acts to keep the interior hot. On top is a thin skin made of the lightest silicate-carbonate rocks at most only a few tens of kilometers thick, the solid *crust*. The lightest parts form the thicker elevated continents on which we live, while the thinnest portions underlie the basins that are filled with ocean water salted with runoff from the land.

The liquid interior iron is, like the outer layer of the Sun, in a state of convection, in which it circulates up and down as well as churns as a result of rotation. The movement of the hot iron creates the Earth's magnetic field, which in part protects us from the solar wind. While cooler than the core, the mantle is still so hot that it is in something of a plastic state. It too then circulates up and down, albeit much more slowly. The uprising flows are so powerful that they have cracked the crust into a batch of individual plates, rather like the way the relentless growth of a tree can fracture a sidewalk.

The part of the rising mantle that breaks through the top cools to form new crust, which shoves the ambient crust aside, and thus pushes the continents around. Some 250 million years ago, Africa and South America were joined (just look at the outlines), as were North America and Europe. Radio telescopes on each continent pointed at the same distant astronomical object can actually measure the separation rate at about an inch (2 to 3 centimeters) per year. Continents crashing together or overriding oceanic crust builds mountains, while the violent crunchings and attendant heat causes earthquakes and volcanoes, which (as in Hawaii) can also arise as mantle plumes from seemingly nowhere.

Figure 2.4 Silvery Moon. While "RIP" may be appropriate, our Moon, seen here in its waxing gibbous phase, has a striking beauty of its own. The dark lunar maria are identified in Figure 2.3. The "mountains" of the Moon are the walls of the massive maria, which are basically huge craters. The lunar highlands to the south are filled with more traditional impact craters jumbled one upon the other. The sunrise line, where shadows are strong and the cratering stands out in bold relief, runs vertically at left; toward the right, under more overhead sunlight, the craters disappear from easy view. A few crater walls to the left of the sunrise line are just catching the first rays of sunlight. Courtesy of Mark Killion.

maria, on the other hand, are younger, generally dating between three and four billion years old. From these and vast amounts of other data, we can put together the path of lunar development.

Origins

The Sun was born 4.5 billion years ago from a collapsing dense cloud of interstellar gas and dust, a concept based on hard evidence derived from the observation of other stars (see Chapter 4). In its earliest days, the Sun was surrounded by a vast dusty disk created as part of its formation. We see such around other young, even older, stars. The planets are widely acknowledged to have been created by accumulation of the disk's dust grains. The inner developing bodies were too hot from sunlight to accumulate the abundant, but lightweight, hydrogen and other substances such as water, and turned into dry rockballs, while the colder outer ones could do so, and grew fat and large. The heat generated by the constant bombardment that created Earth caused a partial melting. The heaviest stuff, the common iron and nickel, sank to the center to make the core, while the lighter rock floated to the outside. The lunar chemistry strongly suggests that it was formed by a collision between the primitive Earth and a smaller competing planet that ran in a crossing orbit. Most of the collider (and its iron) was incorporated within us, while the rest of the debris went into orbit around us and quickly re-accumulated to form the rocky Moon. Evidence for such gigantic collisions is everywhere in the Solar System: even on the Moon itself.

Planet formation was hardly totally efficient. After the planets formed, cooled, and solidified, a lot of small stuff was left over to roam in shifting orbits about the Solar System. Some of it was tossed out of the system by planetary gravity, but a good (and unknown) fraction was swept up by the planets, where it produced a massive heavy cratering episode called the *heavy bombardment*. Within a half billion years or so, it was more or less over.

The Moon was then struck by a series of much larger bodies that produced basins that range into the many hundreds of miles across. Over a period of hundreds of millions of years, seeping lava gradually coated them over to make the maria we see today. The maria that make our familiar Man in the Moon are for the most part really gigantic craters that can be identified by naked eye with no trouble whatsoever, while the lunar mountains ranges, unlike those of Earth, are basin walls. By three billion years ago, most of what we see today was in place.

In the meantime, a much lower rate of impacting has lightly littered the maria with more recent craters, but it was nothing like what happened before. A few big impacts that created the young larger craters such as Tycho and Copernicus hit within the past few hundred million or so years. Some of the impacts were so great that they knocked rock clear off the Moon. A little of it found its way to Earth to become lunar meteorites in what one might call a "cheap space program."

And of Earth?

If the Moon caught it so badly, why did not Earth? It did. But Earth is larger, contains a substantial atmosphere, wind, water, and continental drift that long ago wiped away all the evidence of early cratering. The Moon, however, also displays younger craters. And indeed, so does Earth. A couple hundred of them. Some, like three-quarter-mile wide *meteor crater* in Arizona, are obvious, while others, such as the 150-mile (250 km) diameter Sudbury Structure in Quebec, are hidden to all but deep-rock geologists.

Figure 2.5   Collision! We have impact craters on Earth as well, as evidenced here by 0.75-mile (1.1 km) wide, 50,000-year old Meteor Crater in Arizona, about 200 of them known. Here is one astronomical sight where artificial lighting is of no consequence.

Among the most notorious is Chicxulub. A bit over a hundred miles (170 km) across, discovered during drilling for oil, it lies off the north coast of Mexico's Yucatan Peninsula. Created about 65 million years ago, it coincides in time with the extinction event that ended the reign of dinosaurs. The impactor, thought to have been a seven-mile-wide asteroid, raised so much dust and vapor into the air that the Earth was plunged into a dark, long-lasting winter that killed off much of its life. The initial evidence is a layer around the Earth dated 65 million years ago that is rich in the element iridium, which is far more common in asteroidal meteorites than it is here. The Moon thus provides clues to help us to understand our own home planet.

Where and When: the Phase Cycle

Two lunar phenomena vie for the most attention, the dark markings on the Moon's surface and its continuously changing shape. Among the most pervasive myths — or misconceptions — in astronomy is that the seasons are caused by the changing distance between the Earth and the Sun (they aren't) and that the phases of the Moon are caused by the shadow of the Earth (they aren't either).

The explanation for what we see involves three considerations. The Moon shines only by the reflected light of the Sun. And though it does not look like it, the Sun is 400 times farther than the Moon. Finally, the orbital plane of the Moon is very similar to that of Earth. Tilted by just five degrees to the ecliptic (see Figure 1.4), the lunar orbit and the Moon lie within the constellations of the Zodiac. The phases are then caused strictly by the smoothly changing apparent angle between the two bodies, which allows us to see different portions of the lunar daytime surface and to witness the complementary portions of the nighttime side.

The Moon goes through its set of phases every 29.5 days, which is the origin of our month. In our civil solar calendar, the months are stretched to 30 or 31 days to fit 12 of them into the year. Which is why there are 12 constellations in the western Zodiac. Every time the

Moon starts a new phase cycle, the Sun averages roughly one more zodiacal constellation to the east.

Place the Moon between the Sun and Earth (not quite exactly, so that you do not cover the Sun), as in Figure 2.6. The more distant side faces the Sun and is in full daylight. The side facing us, however, is in night, and you can't see it. This configuration, *new Moon*, starts the phase cycle. Now let the Moon turn a few degrees around the Earth. With the Moon placed a bit to the east of the Sun, we can see a sliver of the daylit side, and the Moon appears as a crescent, as in Figure 2.1. As we watch night after night, we see more and more daylight, the crescent becoming larger, *waxing*. Since, as seen in the sky, the waxing crescent closely follows the Sun, it is visible in the west in late afternoon through early evening, the Moon thus setting shortly after sunset.

At an angle of 90 degrees between the two, we view half the daylit face and half the nighttime side. Though we see half the disk — a *half Moon* — the quarter turn in orbit gives it the more common name of *first quarter*. At a right angle to the left of the Sun (as seen from the northern hemisphere), the first quarter on the average rises at noon, is more or less due south at sunset, and sets around midnight. Bright and bold, it's easily visible in our afternoon daylight hours.

After the quarter, we will now see more of the lunar daytime side and less of nighttime. This *waxing gibbous phase* (Figure 2.4) rises after noon, is easily visible in the daytime, and sets after midnight. Finally, just over 14 days after it started, the Moon is more or less opposite the Sun, and we see the entire sunlit face of the *full Moon* (Figures 2.3 and 2.7). Being 180 degrees from the Sun, the full Moon must rise around sunset, cross to the south at midnight, and set at sunrise.

The cycle thence goes into full reverse, the Moon passing through waning gibbous (when it rises *after* sunset and is visible in daytime morning hours), third quarter (when it is seen again as a "half Moon" that now rises around midnight and sets around noon), waning crescent (when it beautifully shows itself before dawn), and finally back to new. After which the whole cycle starts all over again.

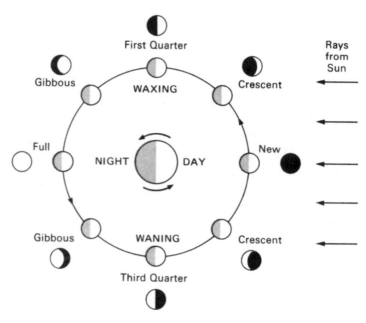

Figure 2.6 The drawing of the lunar phases speaks for itself. The rotating Earth, going through day and night, is in the middle, the Sun *way* off to the right. The inner ring shows the orbiting Moon with shaded night, the outer ring the appearance of the Moon as more and more of the lunar daylight half is revealed to Earthly beings until full, then less and less. Phases correlate with when the Moon is visible. The waxing and waning crescents are seen in evening and morning twilight, first and third quarters rise near noon and midnight, while full phase (opposite the Sun) is up all night. S. P. Wyatt and J. B. Kaler, *Principles of Astronomy: A Short Version*, Boston, Allyn and Bacon, 1974, 1983, copyright J. B. Kaler.

If you think the Moon is bright from the Earth, try the Earth as seen from the Moon. The Moon is made of dark rock, while the Earth is covered by highly reflective clouds and water. From the Moon, the Earth also goes through the same phases as does the Moon from Earth, only in reverse. When the Moon is new, from the Moon, Earth would appear full, and so on. Earthlight on the Moon is so intense that you can easily see it light up the nighttime portion of the crescent Moon, producing an amazingly lovely sight (as in Figure 2.1). By first quarter Moon (third quarter Earth), the effect is gone, only to reappear in the waning crescent phase.

Figure 2.7   The legendary full Moon: poetical inspiration at its best. That it looks so large upon rising or setting is an illusion, a trick of the mind. It's really always but a half-degree across.

Back and Front

Watch the Moon long enough and you see that in spite of its changing shape, the dark markings are always in the same place. The Moon keeps one face — the *nearside* — pointed at Earth, while the back side (the *farside*) is always turned away and never visible. The Moon does not rotate relative to Earth. Slight wobbles allow you to see about 60 percent of it over the course of its orbit, while the other 40 percent remains forever hidden from here. We had no idea what was there until we and the Soviet Union sent spacecraft around the Moon. No surprise, except for the lack of large maria, it looks rather like the nearside, covered with basins and craters, but few ancient lava flows. The Moon *does* however, rotate with respect to the Sun, all portions of it having two weeks of daylight followed by two of night, as evidenced by the phases. Through the telescope, during the waxing phases, we can watch the sunrise line (the *terminator*) creep across the lunar surface, high mountains (really crater and maria walls) catching the first rays of sunlight, while we see the reverse during the waning phases, the lunar features entering night: see Figure 2.4. Because there is no air, there is no twilight, and the terminator is sharp and clear. The origin of this odd behavior lies in the tides, which cause other profound effects that will be tended to later on.

The craters are best seen through the telescope near the terminator where their walls and mountain peaks throw long shadows. At full, the shadows are hidden and the craters become nearly invisible. At full Moon, though, the younger craters' splash marks, the *rays*, become most visible. Under harsh sunlight, they darken with time, their existence allowing you to identify the "new," relatively recent, craters.

Some Oddities Involving the Full Moon

The word "loony" comes from the Moon, Luna in Latin. The full Moon is supposed to bring out odd behavior, criminality, even vampires. The first may be a self-fulfilling wish, crime statistics show that the second one is false, and the third is, well, loony.

But there are others that are fun to follow. The phase cycle of 29.5 days is shorter than the civil month. If full Moon falls on the first of the month, then a month can have two full Moons in it, the second one (for obscure reasons that apparently started with the "Maine Farmer's Almanac" as interpreted by Sky and Telescope magazine) called a "blue Moon." It's not blue. And nobody seems to care about two NEW Moons in a month.

Watch the full Moon rising. It's *huge*, and then seems to become smaller as it lofts itself higher and crosses the sky; see Figure 2.7. It's an illusion, not surprisingly called the "Moon illusion." Really. The Moon's orbit is more or less circular with about a five percent deviation on either side of the average distance. While the Moon does change its angular size, becoming largest at "perigee," smallest at "apogee," the difference is not enough to notice with the naked eye. At Moonrise (Moonset for that matter), the Moon is still just half a degree across, no bigger than it is at its highest crossing at the celestial meridian. And it's really pretty small. As beginning photographers quickly figure out, the Moon is barely visible on straight, non-telephoto images. Indeed, you could circle the 360° sky with 720 edge-to-edge full Moons.

Many cultures, notably many native to America, gave names to the individual full Moons. Rose Moon in June, Thunder Moon in July, Winter Moon in December. Among the best known is the "Harvest Moon" in September (or early October, the one actually closest to the time of the Autumnal Equinox). The delay in Moonrise from one night to the next over the course of the year averages about 50 minutes. But at any given full Moon the actual delay depends on the angle the early-evening ecliptic makes with the horizon. In northern hemisphere September evenings, it is rather flat, and from one night to the next the delay is quite short. As a result, early evening is flooded with near-full moonlight for several days. The October "Hunter's Moon" is similar. You see just the opposite in the spring, when the evening ecliptic is steeply angled, and the waning gibbous disappears very quickly from the early evening sky. Put the Harvest Moon with the Moon illusion, and they combine to make an unforgettable sight, especially if the full Moon is rising in the Earth's shadow cast upon the lower atmosphere and is reddened by haze (see Figure 2.7).

The full Moon has double the apparent reflecting area as the quarters, but is an amazingly eight times as bright! At the quarters, there are strong shadows at the terminator, whereas at full, the shadows are hidden, and we see light reflected from the whole lunar surface. Moreover, sunlight is more efficiently scattered from the Moon's rough regolith in the backward direction than it is to the side, the two phenomena adding together to give us our brilliant full Moon. And if the Sun can produce a blue sky, why not a bright, full or near-full Moon? It does. However, violating our main tenet, sadly not from the city. But it's still there. From a dark clear location, the sky under full moonlight takes on a decidedly bluish tinge.

Calendar Issues

Our civil calendar is strictly solar. It has a legacy from ancient days that incorporated the Moon by being broken down into a dozen months that stretch out the period of the lunar phase cycle. Two other widely-used calendars, however, use the true lunar month. Its tough enough to deal with the irrational fit between the solar year and day, let alone put in yet another non-fitting cycle.

There are, however, popular solutions. Both the Moslem and Jewish calendars in principle use alternate months of 29 and 30 days to get close to the 29.53-day phase cycle. Months are started with the new Moon. Since the "lunar year" of 12 lunar months is just 355 days long, both calendars rapidly get out of synchrony with the solar equinox passages. The two calendars then markedly differ. Falling near the time of the Autumnal Equinox, the Jewish New year and holidays rather quickly move backward relative to the equinox and the civil solar calendar. When they have moved by roughly a phase cycle, the calendar then adds a 13th month, a leap (or "embolismic") month to bring the New Year back to where it was. Various other, more complex, adjustments are needed as well.

By edict of Mohammed, however, the Moslem months stick strictly to the 12-a-year rule. Paying no attention to the equinoxes, they cycle through the entire year. The holy month of Ramadan, which involves fasting during the day, can take place in winter, when it is relatively easy, and then in summer, when fasters long for sunset.

Seasons of the Moon

Over the course of the year, the Sun goes from 23.4 degrees south of the equator at the Winter Solstice (when it is overhead at the Tropic of Capricorn) to 23.4 degrees north (overhead at the Tropic of Cancer) and back again. The Moon, closely following the path of the Sun, behaves much the same way, the main difference being that it goes around in a bit under a twelfth of the time. Check back to Figure 1.4 and look at the main stations of the ecliptic. In September, the Sun is near the Autumnal Equinox, so the Harvest Moon, opposite the Sun, must be near the Vernal Equinox and must therefore rise nearly due east, set due west. The March full Moon, which lies near the Autumnal Equinox, works the same way.

In June, with the Sun near the Summer Solstice, the full Moon is close to the Winter Solstice, and behaves like the Sun does in

December. It rises well to the southeast, crosses low to the south (as seen from the mid-northern hemisphere), is not up for very long, and sets in the southwest. Given summer humidity, a low midnight summer full Moon can be a glorious, rust-colored sight. In December (and neighboring months) we see the opposite. The full Moon, now near the Summer Solstice, rises in the *north*east, sets in the northwest, is up for a longer time, and crosses to the south very high, flooding the winter midnights with a pure white light.

Return to the beginning of spring. Now place the Moon at its first quarter, 90 degrees to the east of the Sun, that is, at the Summer Solstice, where again it rises in the northeast. As fall begins, however, the first quarter lies at the Winter Solstice and rises in the *south*east. Over the course of the year, every aspect of solar motion will be replicated by every phase of the Moon.

The five degree tilt of the lunar orbit against the ecliptic also allows the Moon at some point in its orbit to ride five degrees both to the north and south of the solstices. The extreme "tropics" for the Moon, where it can be seen overhead at some point in its phase cycle, then fall five degrees north and south of the solar tropics, at 28 1/2 degrees north and south of the equator. The lunar version of the Arctic Circle is similarly affected, the limit of the no-setting Moon five degrees to the south of the solar Arctic, or at 61 1/2 degrees north latitude.

Tidal Waters

Among the most profound of lunar effects are ocean tides that make the sea periodically rise and fall at the shore. The explanation is remarkably simple, though the full effects are brutally difficult to calculate accurately. The Moon is held in orbit through its mutual gravitational attraction with Earth. The strength of gravity, as found first by Isaac Newton, drops off rapidly with distance. It is then stronger on the side of Earth facing the Moon than it is at the terrestrial center, and is stronger there than it is at the surface on the more distant side. The Moon thus stretches both the Earth and its waters along a line to itself. The oceans try to flow to that line, but because of the

Figure 2.8   High and low tides at Cutler Maine. The greatest range is produced at new and full Moons, with the Moon at its closest point to the Earth (perigee). Maine Department of Sea and Shore Fisheries.

stretch, toward *both* sides of Earth, the one closer to the Moon as well as the one farther away. As the Earth rotates, it goes under deeper then shallower water, and we see the tides come in, reach a peak, and fall back. High and low tides accordingly occur roughly twice a day (given the lunar motion to the east, each actually separated on average by 12 hours and 26 minutes).

The complications are legion. First, the ocean flow cannot quite catch up with the Moon, so that high tide is always well past the Moon's highest (and beneath the horizon, its lowest) arc in the sky. The timing and height also depend critically on the slope of the ocean bed and the lay of the coast. Coastal tides can range from barely visible to more than 50 feet high, and some are so dangerous as to have drowned the unwary.

Then there is the Sun to contend with. Even were there no Moon, we'd still have tides, compliments of the Sun, which has the same stretching effect, just two-thirds less. The phase cycle thus plays a strong role. At new and full Moon, the solar and lunar tides add together, producing "spring tides" (that have nothing to do with spring), while at the quarters, the solar tide cancels part of the lunar to make much smaller "neap tides." Even the variations in distance between the Moon and Earth and the Earth and Sun get into the act, tides (both high and low) being notably greatest when the Moon is closest. When new or full Moon coincides with perigee the tides are even more enhanced, and when coinciding with a storm, can be disastrous.

The tides have several profound effects. Friction between the waters and ocean bed coupled with the pull of the offset Moon on the tidal bulge slow the Earth's rotation rate and increase the length of the day by about a thousandth of a second per century. To reconcile that with a steadily ticking clock, we must add a second to our time-keeping system about once a year or two, a *leap second*, wherein the last minute of December or June sometimes contains 61 seconds. It thus takes no beach to appreciate the tides. As the Earth slows, the Moon compensates by moving farther from Earth at a (measured!) rate of about an inch a year. Tides raised by the Earth on the Moon — on the solid body itself — have reached their ultimate goal of stopping the lunar rotation completely, so that the Moon keeps one face, the nearside, pointing at us, the farside always away.

In the ultimate effect, some scientists have suggested that life itself started in tide pools, or that tides were at least significant in its origin. In any case, the gravity of the Moon seems to be a stabilizing force in keeping the tilt of the Earths's axis against the orbital perpendicular near 23 1/2 degrees, thereby ensuring the stability of the seasons and, that no matter how life started, that it be maintained.

Dueling Months

The Moon presents a problem that is similar to the one involving sidereal and solar days. The phase cycle of 29.5 days is the period of the lunar orbit relative to the Sun. But the Earth, from which we observe, is moving too, which complicates things. In your mind set the Sun, the new Moon, and a star along the same line of sight and watch them turn about the sky. The Moon moves to the east quickly at a rate of 13.2 degrees per day, its own half-degree diameter in an about an hour, passing through its phases as it goes. That is so fast that if a star happens to be close to the Moon (and there are several bright ones), you can actually see the motion. After a *sidereal period* of 27.3 days, the Moon returns to the star. The Sun, however, has been moving along in the same direction as the Moon at a rate of a hair under one degree (0.99°) per day. By the time the Moon has returned to the star (27.3 days later), the Sun has moved some 27 degrees to the east of where we started. It then takes the Moon (allowing for additional solar motion) another 2.2 days, to catch up with it, resulting in the 29.5-day period of the phases (the *synodic* period, from the Greek for "meeting").

Rings and Things

While the blue sky under bright moonlight cannot be seen from town, most other solar atmospheric effects have their more-easily-visible lunar counterparts (see Chapter 1 and Figure 1.8). Indeed, some are even more frequently seen because there is no danger in looking directly at the Moon. Among the best is the 22-degree halo caused by ice crystals in high clouds. It can come replete with sundog clones that are now called *moondogs* (*or mock moons*), which develop with the Moon when seen not far above the horizon.

By far the most common is a tight halo seen around a fairly bright Moon. Unlike the 22-degree halo, this one is contiguous with the edge of the lunar disk, and the colors are reversed, red on the outside, blue (or white) on the inside. These structures are caused by moonlight shining through transparent water-vapor or icy clouds under a process called *diffraction*, in which light waves passing through a field of tiny water or ice particles interfere with one another, each color building to a peak or valley depending on the angle from the central source. If you wear glasses, you will see diffraction rings around indoor lights when you come in from a cold day and your lenses frost up. The phenomenon is also responsible for the colors that reflect and diffract from the fine grooves of a compact disk. Diffraction halos are even more common around the Sun, but they are so close to the solar disk as to be dangerous to try to look at.

Among the less common is the lunar rainbow, which requires ideal conditions and that once seen is never forgotten. Nor are moonbeams shining through gaps in clouds that seem to provide pathways to the heavens themselves.

Shadows

All bodies in sunlight cast shadows, you included. Children have great fun with them. So must the Earth and Moon cast their own shadows, allowing us all to have great fun. And from almost anywhere. If you have a flat horizon, you can see the shadow of the Earth every clear night as it rises in our atmosphere and every clear dawn when it sets, appearing as a grey band along the horizon right after sunset or before sunrise.

The Earth's shadow extends from the air far out into space. Given the Sun's huge size relative to Earth, the shadow is shaped like a long thin cone that comes to a point at a distance three and a half times that of the Moon. At the lunar distance, the Moon can be completely immersed in it nearly three times over, where it appears in *eclipse*. But since the shadow points away from the Sun, eclipses can take place only when the Moon is opposite the Sun as well, that is, only at full phase.

Figure 2.9 Lunar eclipse. Most of the Moon is immersed in total shadow, while the bright arc at the right still catches some sunlight. The reddish color comes from sunlight thrown into the shadow by the Earth's atmosphere. The shadow's circular outline was one of the first hints that the Earth is spherical. Courtesy of Mark Killion.

Experience shows that we hardly have an eclipse every full Moon. Such *would* be the case if the Moon's orbit lay exactly along the ecliptic. As already pointed out, however, it doesn't. Instead, the lunar orbit is tilted to the ecliptic (the plane of Earth's orbit) by a hair over five degrees. But at the distance of the Moon, the shadow is only about 1.5 degrees in radius. As a result, the full Moon usually passes either above or below the shadow, and we get no eclipse.

However (and there are always howevers), two circles tilted against each other have to intersect in two places. As the ecliptic crosses the celestial equator twice (at the equinoxes), the Moon must cross the ecliptic twice per sidereal month, at the *nodes* (the *ascending* with the Moon going north, the *descending* with it going south). If full Moon happens with the Moon sufficiently close to a node, then we see an eclipse! Very roughly the Moon will average two eclipses a year, once with the Moon near its ascending node, once with it near the descending, though the geometry of the eclipse is such that even with the Moon near a node, the eclipse can still be missed. Given that the Moon is above the horizon half the time, down the other half, any given observer will see half of available lunar eclipses, so maybe you get to witness roughly one (or part of one) per year. Clearly they are not all that uncommon.

There is a funny thing about the lunar orbit, however. The Earth has this third motion, the wobble, or precession, of its rotation axis (see Chapter 1). Here we encounter it again, but this time with the Moon's *orbital* axis, which wobbles with a period of 18.6 years, compliments of the gravity of the Sun. As a result, the nodes move steadily backwards, to the west, opposite the lunar orbital direction, returning to their original positions just under two decades later. If we get a lunar eclipse in early January, the Moon can catch up with the same "regressing" node in December, giving us as many as *three* per year. (The regression of the nodes also allow the *full* Moon to lie five degrees north or south of the ecliptic at some point during the 18.6 year cycle.)

Since the Sun is not a point, any shadow cast under its light has two parts, an *umbra*, within which there is no direct view of the Sun, and a *penumbra*, in which only part of the Sun is cut off. You can see both in your own shadow, which is dark but surrounded by a fuzzy gray band. During a more or less central eclipse, with the full Moon near enough to a node, the Moon first enters the penumbra, moving from right to left against the stars while moving more rapidly across the sky from left to right as a result of the Earth's rotation. Someone on the Moon would then see the Earth cutting off only part of the Sun. The penumbral eclipse is only barely visible from Earth and not usually worth bothering with.

The Moon next enters the umbra, wherein all direct sunlight is cut off. During this *partial* phase of the eclipse, which lasts up to an hour or so, you see a circular cut in the lunar disk that moves progressively across the lunar disk. Finally, the entire Moon is immersed in *totality*. After an hour or so, depending on just where the node is, the Moon then begins to exit the umbra, and the whole thing proceeds in reverse until the Moon is back to its shining normality.

The real wonder of the total phase of the eclipse is that you can still see the Moon! Though direct sunlight is cut off, the Earth's surrounding atmosphere both scatters and refracts it into the shadow, which slightly illuminates the darkened Moon. Somebody on the Moon would look back at our planet and see a ring of light, the bright air, surrounding the "new" Earth. Since air absorbs blue light best, the eclipsed Moon takes on a reddish, almost rusty, color that can be quite beautiful. How much light gets into the shadow, and the brightness of the totally eclipsed Moon, depends on the state of the air, on cloud cover, and particularly on recent volcanic action, which loads the upper air with dust and vapors (and that can produce spectacular sunsets on Earth). The Moon can then range from remarkably bright to almost invisible.

If the node is a bit too far away from the full Moon, the Moon can just miss being fully immersed, and all you see is the partial phase, a *partial eclipse*. If it is farther off yet, the Moon will just pass into the Earth's penumbra, and hardly anything is visible at all unless the Moon is very close to umbral passage.

Eclipses of the Moon are watchable from anywhere, from country to city. The shadow's circular outline long ago demonstrated the sphericity of the Earth to the ancient Greeks. We've know the shape of our home for more than 2000 years. Scientists can also use eclipses to search for meteorite impacts on the lunar surface (as can you with a telescope) and to evaluate the state of the Earth's atmosphere.

Shades of the Moon

Now look at the reverse. If the Moon is near a node at *new* Moon, it can cover the Sun, giving us the gift of a *solar* eclipse. At best, we see the result of the remarkable coincidence of the Moon and Sun having nearly the same angular diameter of half a degree. There are, however, some deep differences with the lunar variety, both in physical circumstances and in nomenclature.

The Moon is a quarter the size of Earth, so its shadow is only a quarter as long. Again a cone, the shadow averages 232,000 miles (374,000 kilometers) in length, with some variation that depends on how far the Earth happens to be from the Sun. The Moon, however, averages 239,000 miles (384,000 km) from Earth, 235,000 miles (378,000 km) from Earth's surface. On the average then, the shadow cone cannot quite reach us. If the centering is exact (and given the diameter of Earth, there is a good positional range), then the observer sees a bright annulus of sunlight; the Moon is a bit too far and cannot quite fit over the Sun, resulting in an *annular eclipse.* While fascinating to watch, it's of rather minimal interest as the bright ring of sunlight obscures the sight that people will travel thousands of miles to see.

Which occurs when the Moon is closer than average and nearer to perigee. The shadow cone can then reach down to us and, with bright sunlight gone from where you stand, the surrounding solar corona emerges from behind the curtain of a darkened blue sky. With the Moon actually at perigee, the full shadow at the Earth at the subsolar point can be nearly 170 miles across (215 kilometers, and larger if elongated near sunset or sunrise), big enough for the Sun to be totally eclipsed for more than seven minutes, giving both spectators and scientists plenty of time to measure and admire it. Except from space, totality provides the only opportunity for us to observe the

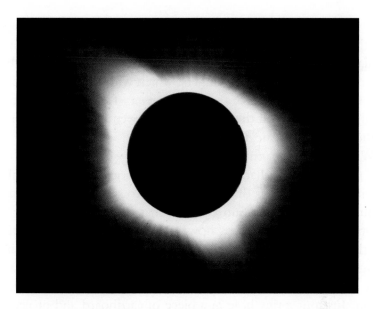

Figure 2.10   Solar eclipse. Now it is we who are in shadow, this time that of the Moon. With bright sunlight cut off, the glorious solar corona (see Figure 1.7) displays itself. At the poles, its gases outline global magnetic field lines, while elsewhere the corona is confined into massive magnetically controlled loops. Naval research laboratory.

properties of the full corona, the importance of which is evident from Chapter 1.

Watching the eclipse, the first thing to be seen is the Moon taking a bite out of the bright solar disk (for the initial partial phase), a sight that put fear into early cultures, and not surprisingly, in some places still does today. Just as the Sun is completely covered, a thin reddish layer that lies between it and the corona, the solar *chromosphere*, pops into view. The chromosphere is a thin red boundary layer that overlies the bright photosphere (see Figure 1.2) in which the temperature first falls from the solar surface and then rises as the corona is approached and that seems to be magnetically heated as well. The first view of the corona flickers in between mountains at the lunar edge, an effect called *Baily's beads*. Reddish *prominences* that seem like flames, but are made of cooler gases, seem to float upward from the chromosphere into the corona, which appears to the eye to be double or more the angular size of the actual Sun. When the Moon uncovers the bright solar disk, the corona briefly lingers, which with the first flash of sunlight makes the Sun into a *diamond ring*. The partial phase then repeats in reverse until the whole affair is over and we must travel to another site perhaps next year.

While the lunar umbra produces only a small speeding shadow spot, the *pen*umbral shadow is very much larger, of continental size, allowing great portions of the population to see a *partial eclipse* with various fractions of the Sun covered that depend on distance from totality. Only during totality, however, can the Sun be safely viewed directly, with no filter. Do not try to make one yourself, as even if dark enough for the eye, it can still pass dangerous infrared or ultraviolet light. A common way of viewing the partial phases is by pinhole projection. Just put a tiny hole in a piece of cardboard and project sunlight onto a piece of paper, or even onto a wall or sidewalk, and you will see a near-perfect image of the Sun, allowing you to follow the eclipse in safety.

Stand under a fully-leafed tree, and the ground is covered with dappled sunlight. The "dapples" are round: they are pinhole images of the Sun caused by sunlight shining through tiny openings between the leaves. During an eclipse, the ground is then covered by small crescents of decreasing or increasing size, depending on how close you are to totality. It's a charming sight that adds to the beauty of the event.

While lunar eclipses are seen by more people — they are visible to everyone in an entire Earthly hemisphere — solar eclipses are actually the more common. The shadow spot is just very small, enough so that any given place on Earth sees a total or annular eclipse on the average of only 300 years. Viewing then requires travel. During an "eclipse season," with full Moon near a node, the Moon can still miss the Earth's shadow and avoid being eclipsed. But the geometry of solar eclipses is such that with the Moon near a node, there *must* be an eclipse of some sort, even if just partial. We can even get two in a row at successive new Moons with a lunar eclipse at the intervening full Moon. Given the nodal regression, there can be, very rarely, as many as five in a year! It last happened in 1935 and won't again until 2160.

Taming the Occult

"Occult," a word with sometimes scary connotations, simply refers to something "hidden." During a total solar eclipse, the Moon occults the Sun. Stars and planets can get the same treatment, their eclipses called *occultations*. As the Moon plies its orbit, it is constantly passing in front of both. While occultations of planets can be pretty to watch, those of stars are of considerable importance. As the leading, darkened edge of a waxing Moon crosses a star, to the eye the star winks out in a veritable instant, the quickness of the event showing that its angular size is exceedingly small.

But with sensitive instruments that have sufficiently rapid response, the time it takes for the star to disappear can be measured. Given the angular speed of the orbiting Moon, which averages 0.5489 degrees per hour, but depends on lunar distance and the gravitational effects of the Sun, we can measure a star's angular diameter. The star's distance (Chapter 5) then gives the physical diameter, which is vital to astrophysicists trying to study stellar natures. Occultations also allow the discovery and examination of double stars whose components are too close to be directly separable through the telescope. It's a tricky business that depends on where the star strikes the lunar disk, even on irregularities at the lunar edge. Among the more wondrous of sights is the *grazing* occultation, during which a star winks in and out of sight at the Moon's edge as it seems to pass between lunar mountains and crater walls.

Given the Moon's angular diameter of half a degree, over the course of a sidereal month the Moon will occult every star within a quarter of a degree of its monthly path. Which, while useful, does not necessarily include a lot of interesting stars. But now once again enter the 5.15 degree orbital tilt, the orbital precession, and the regression of nodes. As the nodes move to the west, the orbit rocks around such that over the regression period of 18.6 years, all stars within

Figure 2.11   Apollo 15 astronaut James Irwin tends to the lunar rover. Mt. Hadley, between Mare Imbrium and Mare Serenitatis, looms over a crunchy landscape beaten up by billions of years of meteoric bombardment. From here we launch ourselves to visit the planets. NASA.

5.4 degrees of the ecliptic (the tilt plus the angular lunar radius) will be occulted, an area of sky that includes a *lot* of stars, including four of first magnitude (Aldebaran, Regulus, Spica, and huge Antares). Slight periodic changes in the tilt angle allow astronomers to squeeze out even a bit more.

Planets can occult stars too. But first we have to be able to identify them and understand their motions.

THREE

WANDERING PLANETS

Venus must have been the first planet I ever saw. Whether they know it or not, it is for most people because of its quite-amazing creamy brilliance. The discovery I best recall, though, is that of Jupiter, which was riveted into my 12-year-old mind. I was watching the Great Square of Pegasus rise. Looking more like a diamond, it pointed downward to this very bright star, which I somehow figured must be Jupiter. The planet's orbital period is 12 years, so I was one Jupiter-year old. I've followed it ever since and have now passed my sixth Jovian birthday. And then there were Mars, Saturn, and finally, of the bright planets, Mercury.

Shift now from singular subjects, the Sun and Moon, to multiple ones, to the Sun's family of nine planets. Or is it eight, Pluto ejected, "plutoed," from planethood. Or if we include our Moon, is it back to nine, or with Pluto 10? Or with other Pluto-sized objects and big asteroids, dozens? The more we learn, the more complex things get. That is, before we know enough to simplify things again.

The essence of a planet (ignoring rising and setting as a result of Earth's rotation) is that it is not fixed in place like the stars, but *moves*, the very word "planet" from Greek "planetai," meaning "wanderers." And their motions are not subtle. Position a planet near a bright star, and the movement is obvious, often over the course of mere days.

Figure 3.1   In a classic scene, the evening crescent visits with Venus, brightest of all planets (at her best 1/1400 as bright as the full Moon). Inside Earth's orbit, Venus is seen only shortly after sunset or before sunrise.

The Moon, going about the Earth, has a simple celestial path that makes it extraordinarily easy to follow. The apparent path of the Sun, reflecting the terrestrial orbit, is even more straightforward. But the planets not only have their own independent paths around the Sun, they are viewed from a moving platform, the orbiting Earth. The *apparent* movements of planets from our perspective then get rather complicated, which disallows simple descriptions of positions with time. The good result is a dazzling nightly parade that never appears the same way twice.

All planets orbit the Sun much as does Earth, in more or less circular paths and in more or less the same planes. All shine by reflected sunlight, their brightnesses as seen from Earth dependent on their distances, physical and thus angular sizes, reflection efficiencies, and various other aspects. Here they are, with average distances, diameters, masses, true orbital periods, and viewing (synodic) periods. Distances are in the fundamental distance unit of astronomy, the

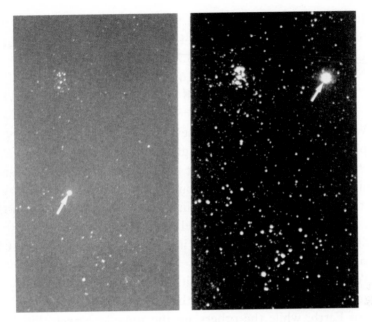

Figure 3.2   Jupiter moves! Over a three-month period (between October of 1988 and January 1989), the Solar System's largest planet, at minus-3rd magnitude second brightest only to Venus, moved nearly 10 degrees westerly — "retrograde" — along the ecliptic against the background of the constellation Taurus. The bright star at bottom is first magnitude Aldebaran, Alpha Tauri (which itself appears against the more distant vee-shaped Hyades Cluster), while the Pleiades (Seven Sisters) lies at top left. The identical images epitomize astronomy's problem. The one on the left was taken from a back yard in a modest-sized city. Obscured by local light pollution, it stands in stark contrast to the one on the right, which was taken from an Arizona mountaintop. Oh what we have lost.

average separation between Earth and Sun, the Astronomical Unit (AU) of 93.2 million miles, or 150 million km.

Mercury (0.39 AU, 0.38 Earth diam., 0.055 Earth mass, 88 d, 116 d)
Venus (0.72, 0.95. 0.815, 225 d, 584 d)
Earth (1.0, 1.0, 1.0, 1.0 yr, ...)
Mars (1.52, 0.53, 0.107, 1.88 yr, 2.13 yr)

Jupiter (5.20, 11.21, 318, 11.9 yr, 399 d)
Saturn (9.55, 9.45, 95.2, 29.4 yr, 378 d)
Uranus (19.2, 4.01, 14.5, 83.7 yr, 370 d)
Neptune (30.1, 3.88, 17.1, 164 yr, 367 d)
Pluto (39.5, 0.18, 0.002, 248 yr, 367 d)

We can divide the planets in several different ways. The first five (skipping Earth), Mercury through Saturn, are the bright *ancient planets*, those known since ancient times, all of which are, or can hit, magnitude zero or brighter, and four of which are, or can be, brighter than the brightest star. The remainder were found in more modern times: Uranus by William Herschel in 1781, Neptune by committee in 1846, and Pluto (let's keep it for old times' sake) by Clyde Tombaugh in 1930.

We can also split them by orbit. Mercury and Venus are the two *inferior planets*, those inside (in the old jargon, "lower than") the orbit of Earth, while the others are the *superior planets*. Finally, we divide them by physical characteristics that have become known only in modern times. From size and mass, the inner four clearly belong together as the *terrestrial* (Earth-like) *planets*, made of rock and iron. The outer four are commonly lumped as the *jovian planets* after Jupiter, though Jupiter and Saturn, filled with gaseous and liquified hydrogen and helium, are really very different from Uranus and Neptune, which are made more from a mush of light chemicals, including a lot of water. The orbital scales of the jovians (illustrated in Figure 3.7) are vastly greater than those of the terrestrials. Outermost Pluto is a ball of rock and ice that is now recognized as the leading member of a debris belt beyond Neptune that is one source of comets (Chapter 6). The bright ancient ones are the focus of our current story, though the outer ones can hardly be ignored (and aren't).

Deep significance. The ancient planets plus the Moon and the Sun are the "seven moving bodies of the sky." As the Moon gave us our month, this septet gave us our seven-day week. In English we recognize (from Latin) the Sun's Day, the Moon's Day, and Saturn's Day, the other four names coming from Norse mythology. Latin and its romantic descendants continue the planetary progression, however, with Tuesday through Friday named after Mars, Mercury, Jupiter, and Venus.

The planets' orbital planes are all close to that of Earth, so the planets are always found near the ecliptic (Figure 1.4). Related to the gods, and "living" in the all-important Zodiac, the planets thus became profoundly important to humanity through their collective astrological role in the prediction of fates. It was not until the astronomers of the Renaissance started to tell us of true planetary natures that the fortune-telling of astrology began to break away from the science of astronomy, allowing us to see the real glory behind it all, which includes the gravitational influences that planets have on Earth, which leads both to collisions with rocks from space and, most likely, to ice ages.

Instant Recognition

Once found, any of the planets can be followed for a lifetime from almost any location on Earth. Each is unique, with a distinctive aspect upon the sky. Planets can be told from stars by their brightnesses and that they move while sticking (mostly) to the constellations of the Zodiac, which string themselves along the ecliptic (see Chapter 1 and Figure 1.4). If it's bright and does not belong there, it's a planet. But such "rules" do not help much on any specific night if you don't have a knowledge of the constellations or if the ambient lights are so bright that you cannot see much in the way of stars in the first place.

You can, though, tell through the old adage that "stars twinkle, planets don't." It's pretty much true. As insubstantial as it seems, Earth's air is a significant refractive medium. As starlight enters the atmosphere, it bends slightly toward the perpendicular. All things in the sky thus appear somewhat higher than they would were we in a vacuum. The effect depends on the angle above the horizon. Near the overhead point it's tiny. But at sunset or sunrise, the Sun is lofted by a full half a degree, which (along with the size of the finite solar disk) advances sunrise by several minutes, delays sunset, makes the Sun a bit oval (Figures 1.1 and 3.8), and places the observable midnight Sun a bit south of the Arctic Circle. Along with refraction goes dispersion, the separation of the light into its component colors.

With magnification, stars near the horizon appear as tiny rainbows, while the clear horizon Sun sports a green rim that lingers above the horizon to make the famed *green flash* sometimes seen in ocean sunsets.

Air, however, is not uniformly layered, but consists of a chaotic mix of cells of higher and lower density and temperature in which the refractive properties differ. An incoming ray of starlight thus bends slightly one way, then the other. As the cells are blown about by winds, a star will appear to jitter around and change slightly in brightness, while the madly changing dispersion causes it to flash in different colors. You can see the same effect in distant street lights.

So far as our vision is concerned, stars are perfect points. But though planets also appear as points to the eye, through the telescope they are not, and instead present significant disks. That is, we can see their surfaces, and often in fine detail. Each actual point on a planet's surface twinkles just as much as does a star (the effect degrading the image). However, the atmospheric cells are small. While one point on the planet is twinkling one way, another may be twinkling differently. The result is that the twinkling on the different parts of a planetary surface averages out, and the planet shines with a (mostly) steady light. As do the Moon and Sun.

But the twinkling test does not aid in actual planetary identification (and no description helps when the planet is not up above the horizon or is hidden by the Sun). We can, however, still recognize them if we know how they look and behave.

Looking for Venus and Mercury?

Venus is easy, Mercury hard. By far the brightest of planets, without question the most striking of them, third brightest of permanent (non-ephemeral) celestial bodies, it's no wonder that Venus is named for the goddess of love and beauty, who comes to us through Latin from Greek Aphrodite. At Venus's best, she hits magnitude −5 (more precisely, −4.7), when she shines 20 times brighter than the brightest star, Sirius (see Figure 3.1).

Coming closer to us than any other planet, 0.28 AU, a mere 26 million miles (42 million kilometers), when the planet is up it's unmistakable. Viewing times, though, are limited. As an inferior planet, it can never be seen very far in angle from the Sun. If it is not so close to the Sun as to be hidden, you'll see it either in the western evening sky during and shortly after the end of twilight, or in the morning before and during dawn, when it takes on the old names of *evening* or *morning star*. Since Venus can't get outside our orbit, except in Arctic climes it's never visible throughout the night. If it's midnight, it isn't Venus. It's *so* bright, though, that if you know where to look, it's readily visible in a clear blue sky in full daylight, while in the deep woods after dark or before dawn it will cast sharp, eerie shadows.

Venus has a 584 day visibility cycle. Start with evening. Going around the Sun, it's first seen as a tiny dot in the west right after sunset, somewhat up from the below-the-horizon Sun. Night after night, it climbs higher and higher relative to the Sun, getting brighter and brighter. If you can see any background stars, Venus will move against them like the Sun, to the east, only faster. It dominates the sky, so much that your local planetarium or astronomy department will get wondering phone calls or emails asking what it is.

It's not a UFO. Venus is the world's favorite Unidentified Flying Object. As you drive home, it follows you. So does the Moon, but most people recognize that for what it is. At night, unlike an airplane, it just seems to hang there staring back. If you happen to see it in daytime and then look away, it "disappears." Put some light evening clouds around Venus and it glows huge with scattered light. Stories are legion, and include a sheriff and his deputy who chased it into the next state and a fighter pilot who tried to shoot it down. The first sight of Jupiter or Mars near their brightest can make people call the cops too.

Some five months after you first spot it at night, Venus reaches its maximum easterly separation (*eastern elongation*) of 47 degrees from the Sun. Though still getting brighter, still going east against the

Zodiac, the Sun now starts to catch up with it. After another five weeks, the planet hits maximum brilliance, appearing awesome against a darkening sky. It then begins to descend seriously back toward the Sun. A fortnight more after maximum brilliance, the motion almost magically reverses as the planet starts to move in the opposite direction, westerly against the stars (a profound characteristic of *all* planets). A couple weeks more and Venus is sadly lost in the solar glare and then passes between us and the Sun at its solar *conjunction* (alignment with the celestial pole), the tilt of the orbit only rarely allowing passage *across* the solar disk.

But wait! Within a couple weeks, Venus pops up in the *morning* sky and then does everything it did in the evening, but in reverse. In just over a month, it's back at maximum brilliance, and then, moving easterly once more (against the stars), it again climbs to its greatest angle relative to the Sun (the greatest *western* elongation of 47 degrees) and then begins a long slow slide of more than half a year back to the Sun. Becoming lost again in twilight, Venus prepares for another solar conjunction, this time more or less in back of the Sun. Then, knowing what will happen, you wait with anticipation until the whole cycle begins again in the west 584 days (the *synodic period* of its meetings with the Sun) after you first saw it.

Odd coincidences, like the similar angular diameters of the Moon and Sun, seem to abound in astronomy. Five Venusian synodic periods of 584 days are almost exactly (within a hundredth of a day) the same as eight Earth orbits, or years. The Venusian visibility cycles then precisely repeat themselves over eight year intervals, making the planet into a long-term calendar. Nobody knows if there is a physical reason for the relationship or not.

As Venus is wonderfully obvious, the other inferior planet, Mercury, is strikingly elusive. Though it can be quite bright, hitting the minus second magnitude (−2.4 at maximum, more than twice as bright as Sirius), its small orbit makes it stick close enough to the Sun that few people ever get to see it, and of those who do, far fewer recognize it. It's said that Copernicus himself never saw it. The little planet, smallest in the planetary system (unless you are a Pluto fan), is *so* close to the Sun that it is visible only in

Figure 3.3   Now it's Mercury's turn to visit Venus, the two seen in near-conjunction (lined up with the celestial pole) hovering in dawn's light. For brief periods of the year, Mercury, on the left (near maximum western elongation), can be amazingly bright: and then it is gone.

twilight. Otherwise, it goes through almost exactly the same cycle as does Venus, only a lot faster, its visibility (or synodic) period just 116 days, allowing for three full cycles over the course of the year.

Given its proximity to the Sun, Mercury is reasonably visible only for a week or so around the times of its maximum elongations of typically just 24 degrees (variations discussed later), and even then only if the angle of the ecliptic to the horizon is steep and favorable for Mercury-watching. It therefore seems to pop up here, disappear, then pop up there, its quickness long ago associating it with the messenger of the gods, the Roman Mercury (Greek Hermes). Yet for all this elusiveness, when it's at its best, the planet is for a few days obvious for those with clear horizons or who live near the tops of tall buildings (or even at the right times fly in airplanes). If you see a bright light not far from the horizon, it sets slowly, and you can rule out the other planets, it's Mercury.

Of Jupiter and Saturn

This pair lies well *out*side Earth's orbit, making these largest two (both truly huge) potentially visible anytime at night. Jupiter is so far away, just over five AU from the Sun, that its distance from Earth does not change very much (from 6 AU to 4), so its brightness does not vary much either. Always at least of the minus second magnitude and at its best reaching into minus three territory (−2.9, nearly four times brighter than Sirius), except for rare moments when it is slightly beaten by Mars, Jupiter ranks number two in apparent planetary luminosity. A visibility cycle of 399 days (just under 13 months) means that if you once find it, it's easy to locate again, as from year-to-year it closely repeats its earlier behavior. And there is little difficulty in recognizing it. Compared with Venus, the planet is clearly off-white, a bit on the yellowish side. If what you see is very bright and not Venus, and especially if it's visible near midnight, it's Jupiter (see Figures 3.2 and 3.5).

Following close to the ecliptic path, the planet visits each constellation of the Zodiac for a year, one at a time, thus taking a dozen years (11.9) to go all the way around the starry sky. Its brilliance and lordly motion surely contributed to its naming after the king of the gods, Zeus in Greek, in Latin Zeus-Pater, literally "god the father." Depending on where it is in its 12-year cycle, you might find it (as viewed from the northern hemisphere) low in Sagittarius or Scorpius near the Winter Solstice, where it rises in the southeast and sets in the southwest, or high in Gemini near the Summer Solstice (rising in the northeast, setting in the northwest, in the middle at the equinoxes, or anywhere in between. (Clearly, Venus and Mercury have the run of the Zodiac too. In the brightness of twilight, though, their positionings are not so obvious.)

Wherever Jupiter might be, we start with the planet invisibly lined up with the Sun, at solar conjunction. Your first view of it will be in the east as it clears dawn's light. Though moving easterly against the stars, it's going slower than the solar motion, so it gradually moves

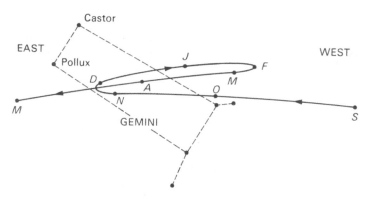

Figure 3.4 All planets exhibit retrograde motion in which they run through a backward (east to west) loop against the stars. Here it's represented by Mars, which begins normal (west to east) motion against Gemini in September ("S") of 1960 at right, then proceeds into the next year, the travels ending in March of '61. Jupiter behaves similarly in Figure 3.2. Mars passed opposition with the Sun in late December, in the middle of its retrograde loop. From *Principles of Astronomy: A Short Version*, S. P. Wyatt and J. B. Kaler, Allyn and Bacon, Boston, 1974, 1983, copyright J. B. Kaler.

away from the Sun, rising ever earlier. Somewhat less than three months after you spot it, it's rising at midnight. Now if Jupiter is near a bright star (one of several within the Zodiac), watch its motion, which gets slower and slower to the east against the star. Around four months after you first saw it near dawn, the motion stops. Jupiter then reverses itself and moves westerly (*retrograde*, much as Venus did, but more obviously).

As another two months go by, the westerly motion speeds up, and nearly six months after first visibility, the planet lies opposite the Sun (the astronomer's *opposition*, a feat impossible for Venus or Mercury), when it rises at sunset, crosses to the south at midnight, sets at sunrise, and is at its brightest. After opposition, we see the reverse. Setting ever earlier, the retrograde motion slows, the planet dims some, and four months past the start of retrograde stops altogether, whereupon the easterly movement resumes. Finally, roughly a year after first sight in the morning, Jupiter disappears into bright evening twilight, setting close after the Sun, finally passing behind it. A bit more than a month later, there it is again in the morning, having found its way to the next zodiacal constellation over to the east.

Saturn does exactly the same thing, just at a different pace. Taking more than twice Jupiter's time to go around through the Zodiac (29.4 years), the planet resides in any one constellation for on the average more than two years. Moving more slowly than Jupiter against the stellar background, the visibility or synodic cycle is also smaller, 378 days. The planet's lagging motion was likely responsible for relating it to Jupiter's father, the defeated titan Saturn (Greek Chronos).

Farthest away from us, Saturn is the faintest of the classical planets and — aside from Mercury — is thus the toughest "find" (see Figure 3.5) Although it can reach magnitude zero, which is pretty substantial, it's still always fainter than the two or three brightest stars, and under the worst circumstances, dips well into first magnitude and is fainter than more than a dozen including some in the Zodiac.

The color, which is somewhat more off-white than Jupiter, the apparent brightness, and lack of twinkle, should still allow recognition. It helps that, like Jupiter, Saturn can be seen at midnight. And there is always elimination, which requires knowledge of the bright first magnitude stars of the Zodiac from Chapter 5. At least once found, the planet moves so slowly that it is easily re-located the following year.

A big help in the recognition of any planet is to know its rising and setting time and the time of its meridian transit (when the planet reaches its highest arc to the south), which are available from a variety of sources that includes the *Astronomical Almanac* (produced jointly by the United States Naval Observatory and Her Majesty's Almanac Office in the UK) and magazines such as *Sky and Telescope* and *Astronomy*. See their Web sites as well. Another is to use predictions of close passages between planets and the Moon. Both are especially useful in the recognition of Saturn and Mercury.

Figure 3.5   In October of 2000, a dozen years after the left-hand photo of Figure 3.2 was taken, Jupiter (the brightest body in the image) again passed through Taurus, but this time was visited by Saturn (second brightest, up and to the right of center). Both planets were in retrograde during a time of their "grand conjunction," which happens (Jupiter passing Saturn) every 20 years. Third brightest, down and to the right of Jupiter, is Aldebaran, which lies about half-way between us the vee-shaped Hyades star cluster.

And Then There's Mars

Mars is unique. As an ancient, superior, terrestrial planet, it's the only one that fits into all three categories. Of all the planets, it has the most distinctive color, a clear orange, making it the obvious candidate to represent the god of war (in Greek, Ares). Just outside Earth's orbital path, the "red planet" has by far the widest range of distances from us and therefore the greatest swing in apparent brightness. It can average as far away as 2.5 AU (235 million miles, 380 million km) and as close as 0.5 AU (48 million miles, 78 million km), a difference of some 500 percent. The planet consequently goes from a relatively dim second magnitude at worst to minus second at average best, making it the fifth brightest of permanent celestial bodies, just after Jupiter. On rare occasion it can even briefly top the giant. So when Mars is near or at its low point, which will be in and near twilight, it can be hard to find, while at other times (near opposition to the Sun and visible at midnight) its brilliance and distinctive color make it impossible to ignore.

Like Jupiter and Saturn, Mars, running the gamut from the deep southern ecliptic to its far northern portion, can be visible at

Figure 3.6   Reddish Mars (to the right of center) stands out in contrast to the fainter and bluish star Dschubba just below it, which lies at the center of the vertical three-star head of Scorpius (the Scorpion). To the left of center, at the Scorpion's heart, is first magnitude red Antares, the "rival of Mars." See Figure 5.2 for a full view of Scorpius.

midnight. However one starts the viewing cycle, the first unimpressive glimpse will be — as it is for much brighter Jupiter and Saturn — in the morning as dawn begins to light the sky. Because of the proximity of its orbit to Earth's (which causes Mars's orbital speed to be just under ours), the Martian viewing cycle is very long, 780 days, or just over two years between successive Martian oppositions to, or conjunctions with, the Sun.

As a result, the early morning view changes with agonizing slowness, Mars coming close to keeping an easterly pace with the Sun. Only after an interval of months will the planet begin to brighten and make its move toward rising significantly earlier. It takes a full three-quarters of a year past solar conjunction before it even rises at midnight. Though the easterly motion has slowed, by then it has brightened so much as to be readily obvious.

Shortly thereafter, Mars's easterly motion halts altogether. Switching gears, it enters westerly retrograde motion (see Figure 3.4), and a little more than a year after conjunction it's opposite the Sun, rising at sunset, crossing the southern meridian at midnight, setting at sunrise. Like Venus and Jupiter, it then makes a conspicuous statement. Just look for the sky's brightest reddish light. But don't get fooled. There are two first magnitude stars in the Zodiac that have similar colors, orange Aldebaran in Taurus and brighter but redder Antares in Scorpius, which can easily look like the planet when it's not near maximum brilliance. The first of these stars rides the northern hemisphere sky high in winter, while the other is low in the south in summer. The possibility of confusion is given by Antares' very name, which means "like" or "rival to" Ares (see Figure 3.6). But at its best, near opposition, Mars far outshines them both. And it hardly twinkles at all. The planet is also speeding in reverse, the motion fast enough that it is actually visible only over a few nights. (Through a high-powered telescope, if the planet is near a reference star, you can almost see it move in real time.)

Outgoing, Mars then just repeats its incoming behavior backwards. After opposition, it slows, re-reverses back to the east, then remains visible for a good part of a year. As the somewhat faster Earth pulls away from it, Mars fades. Hiding in evening twilight for months, it finally and invisibly passes conjunction with the Sun, whence it starts its biannual round all over again.

What Did the Ancients Make of All This?

By the fourth century BC, Aristotle and others knew the Earth was a sphere, in part from its circular shadow during eclipses of the Moon. A century later, Eratosthenes even determined the terrestrial diameter by noting the angle through which the summer noonday Sun seemed to move upon traversing a given north-south distance on Earth.

But clearly the Earth must be at the center of the Universe! Where else would the gods put it? And it's obviously too big to move, to rotate, or to revolve about the Sun. So everything, the entire celestial sphere with its stars, must go about us daily (giving us risings and settings), while at the same time, the Sun, Mon and planets must have additional circular motions around us against the starry background. It's all quite logical, even if almost totally wrong.

The Moon goes around us fairly smoothly (speeding up a bit here, slowing down there), which is the only part that was correct. Hipparchus, he of magnitude and precession fame, even measured its distance with considerable accuracy by cleverly using the size of the Earth's shadow during an eclipse. In the ancient view, the Sun moved even more smoothly about us thanks to Apollo and Helios. But what to make of the planets? Though like the Moon and Sun, they seemed to have orbits around the Earth that generally took them toward the east, every one had these odd reverses in which they went backwards, the inferior planets upon crossing the evening-to-morning conjunction with the Sun, the superior ones looping symmetrically around the times of solar opposition (see Figure 3.4).

Accounting for these movements was not an easy task. But there was more at stake than curiosity. Life and death were involved. The planets epitomized the gods, and to know their dispositions — and our fates — we needed to be able to predict their positions within the constellations — the signs — of the Zodiac. Various schemes were adopted that included nests of offset spheres each carrying a planet, all rotating obliquely against one another.

These Earth-centered *geocentric* ideas culminated in the second century in the *Ptolemaic system*. In the year 150, the Alexandrian Greek astronomer Claudius Ptolemaeus produced his fabled text of astronomy, the *Syntaxis*, which the later scientific Arabs adopted as the *Almagest*, the "greatest book." In his basic system, each planet rode around on a small circular orbit (an *epicycle*) that was centered on a larger geocentric orbit (the *deferent*), which was in turn centered on Earth. With two combined motions, each with a different period, the planets could then make their cyclic loops as viewed from our perspective. Offsetting the Earth slightly from dead center made the scheme work even better, if still not all that well. (Ptolemy and his predecessors had actually invented *Fourier analysis*, in which non-periodic motions are represented by a series of periodic ones.) The concept lasted for 1400 years.

Movements Made Simple

Then a Polish cleric named Nicolaus Copernicus took it away. His Sun-centered *heliocentric* ideas were preceded by those of ancient Greek iconoclasts, notably Pythagoras (who championed a rotating Earth) and Aristarchus of Samos (who in the fourth century BC actually tried measuring the distance to the Sun). After a lifetime of work, in 1546 he published his *de Revolutionibus Orbium Celestium* ("On the Revolutions of the Celestial Spheres", the last two words of the Latin title almost always dropped). In it, he described in detail, with appropriate mathematics, how much more sense could be made of the appearance of celestial motions if the Earth rotated on an axis while at the same time everything except the stationary sphere of stars went about the Sun in the counterclockwise direction as viewed from above the north pole (and thus generally to the east). From their motions and positions, he even determined the distances of the planets from the Sun relative to the Earth's distance, the AU.

Look at the beauty and simplicity of it all. Like the Moon, each planet has two periods, *sidereal and synodic*. The sidereal period is the actual time it takes the planet to revolve from one point in orbit — or from one particular star — and back, while the synodic (related to the visibility cycle) is the interval between successive conjunctions with (or oppositions to) the Sun. The more distant a planet is from the Sun, the bigger its orbit and the farther it must go. In addition, more distant planets move more slowly. As a result, sidereal periods rapidly increase with distance, as seen in the introduction to this chapter.

Jupiter, with a sidereal period of just under 12 years, sets a fine example. Start with it at opposition to the Sun, say in Pisces, Sun-Earth-Jupiter all in line (Figure 3.7). While the Earth zips about the Sun in a year, Jupiter moves along through just a twelfth of its much longer path. By the time the Earth has returned to its starting position, Jupiter has moved 30 degrees to the east, one constellation over, into Aries, and it takes the Earth another month to catch up to it, to bring it back into opposition with the Sun. The synodic period of Jupiter, which is no more than the lapping time of the planet by the Earth, is thus 13 months (more exactly, 399 days). Slower moving Saturn gets lapped faster, so its synodic period is less, 378 days. Those of the far outer planets approach a year as they get more distant.

Now start with the Earth catching up to a superior planet. As speedier Earth starts to pass it, it just looks as if the planet is going backwards, with maximum angular speed taking place at opposition to the Sun (as in Figure 3.4). To a faster runner on the track, anyone he or she passes just looks like the competitor is going in reverse. It's no more complex than that. But only if you concede that the Earth is moving too.

The inferior planets have no oppositions to the Sun. Instead, they go through two different kinds of conjunctions, *inferior* when the planet passes between us and the Sun, *superior* when it goes around in back, on the other side. From our perspective, Venus and Mercury seem to go into reverse when they swing through inferior conjunction. If you were on Venus, when Venus goes into retrograde from here, the Earth would appear in retrograde from there. Watch them. They really do move.

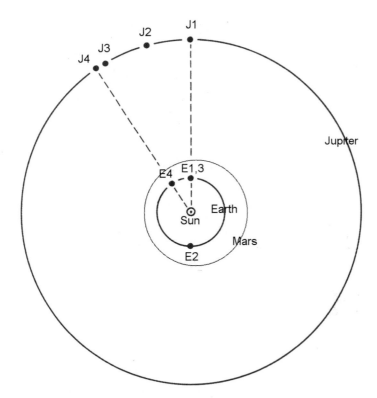

Figure 3.7   The orbits of the giant planets, Jupiter through Neptune, are much larger than those of the terrestrials, which are illustrated by Earth and Mars. Each planet (excluding Earth) has two periods. The sidereal period is the time it takes to go all the way around the central Sun (small circle with dot). The synodic period is the time between successive oppositions to the Sun as seen from Earth. At position *J1*, Jupiter stands in solar opposition. By the time Earth has completed one annual revolution through position *E2* back to *E3*, Jupiter has moved from *J1* through *J2* to *J3*. It then takes the Earth just over another month (with Jupiter constantly moving) to bring Jupiter into opposition again at *J4*, which gives it a synodic period of 399 days. Note the eccentricity of the Martian orbit. On the average, the planet is 75 percent farther from Earth's orbit when at aphelion than when it is at perihelion. Adapted from *The Ever-Changing Sky*, J. B. Kaler, Cambridge University Press, Cambridge, 1996.

Visitations

The planets do not just visit the Sun during conjunctions, they visit with each other and with the Moon as well. Given that the Moon moves much faster around the sky through the Zodiac than any of the planets, Moon-planet conjunctions, some with close visitations, are monthly occurrences for each one. There is little more charming than seeing Venus in a close meeting with a crescent Moon, the only phase during which a conjunction is possible (see Figure 3.1). It's a symbol so powerful that it appears on national flags. Though the superior planets are not as bright as Venus, and their conjunctions not as dramatic, they are still quite pretty. Moreover, such conjunctions can take place at any phase, any of the planets usually able to hold their own even when the Moon is full. Moon-planet conjunctions provide a fine avenue for planet recognition. Though rather rare because of different tilts against the ecliptic, a high point occurs when the Moon actually occults a planet, as described in Chapter 2.

Planet-planet conjunctions are frequent occurrences as well, that of Jupiter and Venus usually making the news (again allowing planet recognition), while the color contrasts in conjunctions between Mars and the others can be quite lovely. Conjunctions between Venus and Mercury (Figure 3.3) can give the eerie feeling of being stared at by two disproportionate celestial eyes. The big one, though, is the long-term *grand conjunction* between Jupiter and Saturn. Given their respective sidereal periods of 11.9 and 29.4 years, it takes Jupiter almost exactly 20 years after passing Saturn to go around the sky and catch up with it again. During a grand conjunction season, retrograde motions can cause the two planets to pass each other up to three times. The last set took place during 2000–2001 (Figure 3.5). Given good fortune, over a lifetime, one can witness three or four of them.

And of course the planets are constantly passing by the Zodiac's first magnitude stars: Taurus's Aldebaran (Figures 3.2 and 3.5), Gemini's Pollux (plus Castor; Figure 4.2), Leo's Regulus (Figure 5.15), Virgo's Spica (Figure 5.1), and Scorpius's Antares (Figure 3.6), often with nice color contrasts. Such juxtapositions allow quick identification of both.

Follow the Rules

As they say in school. The planets do it as well. Indeed, they have no choice. Copernicus had a problem in that he believed from Greek philosophy that planetary orbits had to be circles. He was wrong. As a result, predictions of position were no better than what Ptolemy could do. Revision was called for. The result was three rules of planetary motion, each told twice.

Science progresses as much by accident as it does by following methodology. Thanks to a star that exploded brilliantly in 1572 (Chapter 7), the Danish nobleman Tycho Brahe became interested in astronomy. Establishing a pre-telescopic observatory on a Danish island, Tycho deftly measured the positions of stars (in a celestial coordinate grid akin to latitude and longitude) and, more important, tracked the positions of the ancient planets. Johannes Kepler, who worked with Tycho before the latter's death in 1601, worked backward from the observations to find that a planetary orbit was not a circle at all, but had to be an ellipse with the Sun not at the center, but offset to one focus.

The Ellipse. Place two tacks on a board with a loose string tied to them. Draw a curve with the string tightly stretched. The curve is an ellipse, the tacks the two foci. Bring the tacks together and you get a circle, pull them apart and the curve flattens. The size of the ellipse is given by half that of the major axis, which is the longest line that can be drawn within the closed curve, while the "eccentricity" is the degree of flattening, which starts at 0 for a circle and approaches 1 as the ellipse approaches a flat line.

Planets, including Earth, must then continuously change their distances from the Sun as they go through perihelion and aphelion. The Moon does the same relative to Earth. The quoted distance of a planet from the Sun is an average that equals the ellipse's formal size (half the length of its major axis). As a planet moves against the stellar background, the changing distance and direction cause variations in its apparent angular rate of speed. But there is more to it than that.

To account for all the variation, Kepler had also to make a planet move faster as it approached the Sun, slower as it receded, with maximum speed at perihelion, minimum at aphelion (all precisely defined). With these two rules, in one of the most stunning advances in science, he replicated Tycho's positional observations of Mars.

That was in 1609. It had taken him nearly a decade of work. Together the two rules explain the Greeks' inequality of the seasons (that northern winter is shorter than summer), explain why the date of earliest sunset and latest sunrise do not coincide with the shortest day, show one reason why we need mean solar time, and even factor into monthly and annual variations in the tides.

But Copernicus was not done. After another decade, he had worked out the relations among planetary orbits. In his *harmonic law*, he showed that the square of the period of a planetary orbit (the number multiplied by itself) expressed in years is equal to the orbital size in AU cubed (the number multiplied by itself twice). If you know how far away a planet is from the Sun, or a spacecraft for that matter, you also know how long it is going to take to go around.

But that is not the end of the story. Kepler had no idea *why* the rules applied. But Isaac Newton *did*. Toward the end of the seventeenth century, he discovered that his law of gravity (Chapter 2) led directly to the three rules, but in a much more elegant form that applied to two orbiting bodies anywhere, not just to the planets going about their Sun. Newton's version of the first law allowed open-ended, one-way, orbits in which a body comes in, zooms around the Earth, and goes back out never to return. His second law just reflected the fact that when a planet gets closer to the Sun, it falls faster, in part as a result of being in a stronger gravity field. The biggest change is that Newton's version of Kepler's third rule contains the sum of the masses of the two bodies, which was absent from Kepler's original because the sum of the solar mass plus that of any planet is essentially the same for all of them. Application of Newton's brilliantly-conceived form allows calculation of the Sun's mass. Or of Jupiter's through analysis of the orbits of its moons. Or of Mercury during a one-way flyby of a spacecraft. And so on. Even the masses of pairs of stars.

Crescent of Venus

As Kepler was working out his rules, Galileo was sweeping the sky with the first astronomical telescope. He found moons going around Jupiter, which he took as proof that the planets could orbit the Sun. He discovered craters on the Moon, implying that it was rather like Earth. Sunspots revealed that the Sun was not prefect as then believed. He and his telescope also unveiled the truth about the Milky Way, that it was made of lots of stars, which opened the study of our Galaxy.

But the prize was probably Venus. At superior conjunction, we look across the Sun to the opposite side of the Venusian orbit to see the planet's full sunlit face. (In principle only; don't try it as the Sun is too bright.) Venus is then, like the Moon, "full," with its nighttime side turned away from us. At inferior conjunction, however, with Venus more or less between us and the Sun, it's the sunny side that is turned away. The planet now shows us the Venusian night, and it is "new." At greatest elongations Galileo could see half the daylight face, Venus appearing as a half Moon does in one of its orbital quarters. Between full and half, Venus was gibbous, between half and new, a crescent. You can't get all the phases in a Ptolemaic system. Copernicus 1, Ptolemy 0. Even a low-power telescope shows the phases. You too can be Galileo. Mercury of course shows them as well. Even the superior planets (Mars especially) have slight gibbous phases unless viewed face on at opposition.

While the phases cannot be seen with the naked eye, once we know they are there, we can still note their effect on Venus's brightness variations, which depend on the amount of illuminated *angular* area we see from Earth (expressed in fractions of a square degree). While always bright, unlike the Moon, Venus is faintest around the time of superior conjunction, when it is on the other side of the Sun. Though at or near full phase, it's far away, appears small, and reflects comparatively little light to us. As night after night it climbs the sky, the waning of the phase is more than offset by Venus's growing closeness and increasing angular dimension. Even as a crescent following greatest eastern elongation, the proximity effect is dominant until just before the onset of retrograde, when the planet is angularly so large that the now-very-thin crescent is almost visible to the naked eye. (There are reports of sharp-eyed youngsters seeing it. When Venus is at her brightest, try binoculars.) Then the whole affair is repeated backwards in morning skies.

"Eclipses"

The Moon's orbit is inclined by five degrees to the plane of the ecliptic, and so usually passes above or below the Earth's shadow at full phase, or above or below the Sun at new, in both cases avoiding eclipses. Likewise, all the planets have their own orbital tilts of a few degrees that make them appear a bit either above or below the ecliptic, which they cross twice every sidereal period at their nodes. Mercury's tilt is 7°, Venus's 3.4°. But the Sun is only a quarter of a degree in angular radius. So when Venus or Mercury go through their inferior conjunctions, they too usually pass above or below the Sun, missing it by a wide margin.

But the same rules apply as for the Moon. If inferior conjunction occurs with the planet sufficiently close to a node, it will cross directly against the face of the Sun in a *transit* that is akin to an annular eclipse of the Sun by the Moon. The amount of sunlight cut off is so small that it cannot be detected by eye, but through a properly filtered telescope, one can watch a black dot move against the solar disk, a "sunspot in motion." Only then can you see an inferior planet in its exact new phase.

Transits of Mercury are fairly common. They take place either in November and May in cycles of 7 and 14 years (the November events occurring twice as often as those in May) and require a telescope to see. By contrast, transits of Venus are rare events, occurring two at a time in June or December each eight years apart, each monthly set separated by 243 years. The last December pair took place in 1874 and 1882, with the June pair taking place in 2004 and 2012. The twentieth century lost out altogether. Larger and closer than Mercury, transiting Venus is a marvelous sight that is visible in (properly filtered) binoculars.

Figure 3.8 Transiting the Sun in June of 2004, Venus appears as a tiny dot in the lower right quadrant of the solar disk, vividly revealing how small the planets really are. Refraction by the Earth's atmosphere squashes the solar disk, while the increased thickness of the air through which we look causes the increased reddening and dimming toward the horizon. See the last section of Chapter 1 and Figure 1.1.

Timings of both kinds of transit were historically important in establishing longitude on Earth. The most significant feature of all, though, is being able to watch a planet move in real time. Which you or your descendants can do for Venus in 2117 and 2125.

Eccentricity

Like people, some planets are non-conformists. Most of the planetary orbits are close to circular. The Earth's distance from the Sun, for example, varies by just 1.7 percent from the average. There are three big exceptions: Mars (9 percent), Mercury (21), and Pluto (a whopping 25). Each is highly significant: as is that of Earth.

Though Venus enormously changes her distance from us over the course of its synodic cycle, the brightness variation is smoothed over by the phases. So is Mercury's. Not so of Mars, which displays a very large luminosity difference between solar conjunction and opposition. The god of war, though, has another trick that affects his visibility.

The large eccentricity, or oblateness, of the orbit produces a big variation in the distance at opposition to the Sun. At the average opposition, Mars is 49 million miles (78 million kilometers) away. However, if opposition occurs at Martian aphelion, the planet is 62 million miles distant, while if at perihelion it lies just over 35 million miles away, a swing of some 75 percent, which produces a variation of a factor of three — over a full magnitude — in apparent brightness (see Figure 3.7). Given the 2.1 year synodic period, Mars then undergoes a secondary cycle in which the best (*favorable*) oppositions take place every 15 to 17 years. At the extreme favorable opposition (factoring in the eccentricity of Earth's orbit as well) Mars can top Jupiter in brightness.

Given that at favorable opposition the planet is not only then near its brightest, but that it appears as large as possible in the telescope, Mars becomes a media event with people swarming observatory open houses to get a look. At the last very favorable opposition in 2006, an unkillable rumor that Mars was going to be "as big as the Moon" made the rounds of the Web, and is still alive and well. The "Great Mars Hoax" began with a mis-interpreted news note that the telescopic view of Mars would be comparable with the naked-eye view of the Moon.

Mercury has not one, but *two* eccentric tales to tell. Its large orbital flattening makes its greatest elongations vary between 19 and 28 degrees, which strongly affects the planet's visibility from Earth and makes it a better sight from the southern hemisphere. Of vastly more significance, planets do not live in isolation from one another. Each has a gravitational pull on the other that slowly but inexorably alters planetary orbits. As a result, Mercury's perihelion should move forward around the Sun at a rate of 0.148 degrees per century. The observed rate, however, is 0.159 degrees per century.

While the difference may at first seem trivial, nineteenth and early twentieth century astronomers could not make it go away. The solution had to wait for 1916 and Einstein's relativity. Newton's theory of gravity, as expressed in Chapter 2, is actually slightly wrong.

In Newton's theory, there are no restrictions on speed, while in the real world, relative speeds are limited to that of light. Einstein also viewed the Universe as a four-dimensional structure in which the three spatial dimensions were integrated with that of time. As you approach light-speed, the spacetime world becomes increasingly distorted, making masses increase and slowing time. In relativity, gravity is caused by a similar distortion, a bending, of spacetime by mass. (*Why* this happens, nobody knows.) Using the complex relativistic equations resolves Mercury's problem. It's far more than academic. The relativistic view must be used for spaceflight, including the satellites of the popular *Global Positioning System*, the *GPS*. And Mercury helped lead the way.

Then there is Pluto. Admittedly, at a typical 15th magnitude (nearly half a million times fainter than first magnitude and 4000 times fainter than the human eye can see), it is not exactly one of the sky's bright lights. But as we'll see in Chapter 5, it oddly relates to them. For now, just note its big orbital eccentricity, which carries it between 49 and 30 AU from the Sun. Near perihelion, there is a 20-year period in which Pluto actually comes closer to us than Neptune. That, an orbital tilt of 17 degrees that can take it well out of the Zodiac, and an orbital locking with Neptune (Pluto orbiting the Sun twice for every three orbits of Neptune) quite clearly show that it does not belong with the rest of the planets. Discovery of thousands of other icy rockballs in and outside of its orbit reveal that Pluto is among the largest of a set of distant orbiting debris that lies in a ring beyond Neptune called the *Kuiper Belt*.

And finally, there is Earth. While its orbital oblateness is now modest, aphelion distance being just 3.5 percent farther perihelion's, it was not, nor will it be, always so. Calculation shows that gravitational effects by the other planets can make the variation change over a 100,000-year period from as small as zero to as large as 12 percent, which in turn causes an annual solar heating variation of some 25 percent. The eccentricity variations coupled with the precession (wobble) of the Earth's rotational axis, and a long-term variation in the axial tilt, together called the Milankovitch Cycles, are widely believed to be the triggers for the onset of ice ages.

The Real Thing

Our views of the planets can only be enhanced by knowing something of their real natures. It's rather akin to hearing a symphony being played while reading the score. Our knowledge comes from a combination of Earth- and space-based observations, radar, and through direct visitation by spacecraft. Only Pluto has not yet been so visited, though a spaceship is on its way. Take the tour, from inside to out. While visiting, especially note the relations between planetary natures and solar distance.

Mercury

And now for Channel 1000's planetary weather forecast. Chilly at sunrise. Noontime temperatures will approach 800°F under a hot Sun. There's no windchill though, since there's no air, and no clouds either to hide the Sun. With full radiative cooling, tonight's temperatures are expected to be 300° below, so "cover up the tomatoes."

Mercury is a nightmare. Aside from poor rejected Pluto, it's the smallest of planets. The view from here is also pathetically bad, always in twilight through a thickened atmosphere. The Hubble Space Telescope cannot view it as it is too close to the Sun. We must instead rely on radar and direct visitation. Only 40 percent our distance from the Sun, the average noontime surface temperature climbs to 430 C (800°F), hot enough to melt lead, and it's even hotter when Mercury is at perihelion. There is no air temperature as such, because there's no air, the high heat and low gravity unable to keep any. What little gas there is made of heavier atoms — mostly potassium and sodium — kicked off the surface by the high-speed impacting solar wind. Nor is there any water except perhaps for some ice hidden in deep, perpetually dark craters at the rotation poles.

No water, no air, no erosion. Mercury, like the Moon beaten up by a heavy bombardment, displays a heavily cratered surface as well as huge basins set into smoother plains. The impacts that caused the biggest basins must have come close to cracking the planet in two, giving credence to the theory of the formation of Earth's Moon. The smallest planet being nearest the Sun is no coincidence, and is perhaps related to the lack of raw material in the planet-forming disk, aided by great solar heat preventing its condensation. Collisions may also have stripped off or prevented the formation of the outer rocky mantle, leaving Mercury with a relatively huge iron core that contains 60 percent of the planet's mass, double Earth's ratio.

With no atmosphere, nighttime temperatures fall to −185 C (−300°F). But you'll have a long wait to get out of the heat of the day. Rather like Moon to Earth, Mercury is tidally locked onto the Sun, but oddly. It rotates relative to the stars with a period exactly two-thirds of its sidereal orbital period. On the planet, a full day,

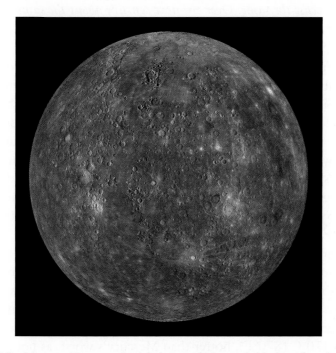

Figure 3.9   Mercury Messenger's global view of Mercury's bleak surface shows it to be covered with myriad ancient craters akin to those on the Moon plus larger basins, all tangled into smooth volcanic plains. Here and there are a few young rayed craters from more recent impacts. NASA/Johns Hopkins Applied Physics Lab/Carnegie Inst. of Washington.

sunrise to sunrise, is 176 Earth-days long, double its orbital period. Given the high eccentricity and resulting variation in orbital speed, there are places on the planet where the Sun rises, slowly sets, and then rises for the second time. Best not to go to watch. Even for visiting spacecraft, conditions are brutal.

Venus

Sunrise at 6 am. Today's temperature about 890°F, so keep your air conditioners on. Cloudy all day with winds from the east about 6 miles per hour. Tonight, sunset at 6 pm. Cloudy with temperature about 890°F, winds from the south at 6 mph. Acid rain will evaporate before it hits the ground. Tomorrow, the same. Five day outlook the same. Long-range forecast the same. Over the next century about the same. Tune in later for an update.

Venus is not a lot better, and for visitors may in fact be worse. It's remarkably similar to Earth in mass and radius. Just 30 percent closer to the Sun than we, old ideas had it as a tropical paradise. Not that one could tell, since the surface is perpetually hidden by clouds, which are one of the factors in making the planet so very bright. Almost all we know about the it has come from cloud-penetrating radar aboard visiting spacecraft. The clouds, made largely of sulfuric acid droplets, float high in an extremely thick carbon dioxide atmosphere that presses on the Venusian surface with a crushing pressure 100 times that of Earth's. It's a lethal place.

Planetary surfaces are heated by sunlight and cooled by re-radiation of infrared light. Carbon dioxide and water vapor are fierce infrared absorbers and act like blankets to keep the Earth warm (the *greenhouse effect*). Without it, our oceans would freeze. On Venus, CO_2's greenhouse effect has gone wild, driving the surface temperature to 470 C (870°F), hotter than Mercury's worst, as hot as a self-cleaning oven. In spite of a ponderous rotation of 117 Earth days relative to the Sun, the thick air allows for no cooling over the 58-Earth-day Venusian night. The heat has driven off all the water, sunlight breaking it down to hydrogen and oxygen, which has then

Figure 3.10 In contrast to Mercury, Venus is covered with turbulent acidic clouds, making it impossible to see the solid surface, which (from radar observations) is covered with relatively young volcanos and impact craters. Venus Express: ESA/MPS/DLR/IDA.

combined with sulfur and what hydrogen is left to make the acid clouds. With no moisture, carbon is not tied up in rocks, leaving it all for the atmosphere. The greenhouse effect has simply run away with itself.

Surface rocks softened by the high heat seem to have contributed to an enormous number of volcanoes akin to ones that make the Hawaiian Islands. Between them, the surface is littered with large impact craters. Given their number and the known rate of impacts from studies of our Moon, the surface seems to be everywhere no more than a billion or so years old. The planet must have undergone a global volcanic flood that destroyed all that was there before and that helped cool the core, leaving Venus magnetically dead, with little protection against the solar wind and not much internal heat with

which to drive any sort of continental drift. Indeed, there are no continents as such, just volcanic rises. And basins with no oceans.

Soviet landers reveal a bleak plain made of flat surface rocks.
Surface winds blow softly. The tilt of the Venusian rotation axis is near
zero, so there are no seasons. Venus not only rotates slowly, it spins
backwards. The Sun always rises almost exactly in the west at 6 AM
local time, sets in the east 12 hours later, each Venusian hour nearly
five Earth-days long. But because of the perpetual clouds, you'd not
know the source of the dim light. Nor would you ever see the stars
until your Venusian NASA sent rockets upward to take a look.

Mars

*Today, wind from the south at a few miles an hour. Clear. Noon surface
temperatures about zero degrees F. Tonight, full radiational cooling with
dropping temperatures. There will be some dry ice precipitation and
watch out for slick water frost in the morning. Weather warning for the
next few months: planet-wide dust storms and winds to 100 mph will
shut down all planetary services. Stay tuned for solar wind warnings.*

Venus and Mercury are "terrestrial" in their overall constructions,
with metal cores and rocky mantles, but they are far different from
our Earth. No one could reasonably live there. Intriguingly, Mars is
not so different from our own planet. In fact it's downright inviting.
From here we see an earthlike rotation of just over 24 hours, polar
caps, an axial tilt close to Earth's that gives Mars seasons remarkably
similar to ours, seasonal snows, dust storms, and variations in the size
and color of its dark markings. It even has a pair of moons, though
they (Phobos and Deimos) are little more than big rocks just a few
miles across.

At one time, various astronomers thought they saw a network of
fine lines across the reddish surface, while the numerous dark markings waxed and waned with the seasons. The lines, so some thought,
must be canals, artificial channels to bring polar water to the temperate farms! Sadly, no. There are no Martians. An armada of spacecraft
has shown that with the exceptions of some natural features, the

Figure 3.11   While no single picture can capture the varieties of Mars, this sweeping panorama near a large impact crater taken by the Opportunity Rover does as well as any, the scene looking remarkably like an Earthly desert. It's hard to believe that water once flowed freely on this planet. So where's the life? Mars Exploration Rover Mission, Cornell, JPL, NASA.

canals are optical illusions, while the seasonal variations are caused by wind-blown dust.

But what we *do* find has perhaps even more fascination. The Martian air is light, just one percent Earth's, and like Venus's is nearly all carbon dioxide. Oxygen is bound up in the Martian "soil," the regolith, its red color coming from iron oxides: rust. Far from the Sun, with little air to trap heat, the temperatures stay low, −30 C (−22°F) under a summer Sun, ground temperatures though hitting close to freezing. At night, temperatures quickly fall to −100 C (−150°F), the thin air not providing much insulation. During the winter the relevant polar cap grows much as does Earth's. The impermanent snows, though, are really no more than dry-ice (frozen carbon dioxide) frosts.

While the thick polar caps are a mixture of dry and water ice, the low air pressure prevents any liquid water. Yet aerial examination by spacecraft reveals networks of dry valleys in the southern hemisphere and huge channels in the northern plains in which water once vigor-ously flowed, which tell of a far more Earthlike planet with a once-thick atmosphere. Orbiting and landed spacecraft support a one-time watery nature. But it must have been a long time ago, as heavy cratering in the south tells of an ancient surface comparable with that of the Moon.

While planets increase in size from the Sun to the Earth, Mars turned out smaller, presumably the result of the effect of Jupiter's gravity disturbing the raw material out of which the planet was made. Martian gravity is low, just 40 percent Earth's. Mars is just not big enough to hold on to much protective air, hence water, which explains the low erosion of southern surface features.

But while the south is old, the north is younger and covered with volcanic rises and the remnants of monster volcanoes. The biggest (Mount Olympus), large enough to cover New England, towers more than twice the height of Everest above what would be — were there oceans — sea level. (But there may have been once, as evidenced by ancient shore lines.) Running east to west across a quarter of the planet is a huge complex crack, the Mariner Valley (named after one of the early spacecraft), which must have been caused by early geo-logic activity, an attempt at continent building that went bad as a result of the quick cooling of the interior. With no continental drift, the volcanoes could grow vast, and in the lower Martian gravity, far taller than anything we have on Earth.

There is a great body of evidence that says the water, which appar-ently covered the planet in abundance, is now frozen below-ground, most of it in arctic or near-arctic latitudes. It will come in more than handy if we ever get there, a daunting journey, one that will take months of exposure to raw interplanetary space with its harsh radia-tion and the solar wind. And there seems little hope of Martians of any sort, including microbic ones. Remote testing reveals nothing. If life is or was there, it is certainly not jumping out at us. And today, with no liquid water, with no air protecting the ground from solar radiation, with the solar wind impacting directly on the planetary surface (thanks to the absence of a significant magnetic field), it would be hard to see how it could survive. But if there is none, why not? There is ample evidence that the planet was once far more earthlike than it is today. What then brought life to Earth, and not to Mars? Life or not, either way, we learn something special.

Whatever the turbulent history of the planet may have been, the current scene is mostly quiet and peaceful. Landers and rovers reveal endless stretches of the reddish dusty Martian regolith littered with boulders flung from nearby impact sites. Except during planet-wide seasonal dust storms, winds blow quietly, while thin water and dry ice clouds form over high peaks. On one side of us is Venus, far too hot for comfort, on the other side Mars, now far too cold and dry. And here we are in the middle, where it is just right. With luck and care, it will stay that way.

Jupiter and Saturn

Travel forecast: Cloudy today with temperatures hovering around −240°F for Jupiter, −290°F for Saturn, near the cloud tops. It will be a lot warmer inside. Winds on Jupiter will hit a couple hundred miles per hour, Saturn five times as much. As expected, Jupiter's big southern storm has not abated, while Saturn's is being spread around the southern hemisphere by winds. Watch for occasional powerful lightning storms in the north. Jupiter's magnetic forecast: still lethal. Expect strong volcanic action on Jupiter's inner big moon Io and passing cold methane showers on Saturn's Titan.

Though the four outer planets are often lumped together as "jovians," there are really only two of them, Jupiter and Saturn. While there are important distinctions between them, pretty much what you say about one is good for the other. First, unlike the terrestrials, which are all roughly comparable to Earth, J&S are *big*, hugely reversing the trend toward smaller planets that began with Mars and ends with a debris belt of asteroids outside the Martian orbit that will be featured in Chapter 6. Jupiter is 11 times Earth's diameter, and carries just over 300 times its mass: a thousandth that of the Sun. Saturn, with 30 percent Jupiter's mass while almost as big (9.5 Earth diameters) once again begins a descent in size that is continued through Uranus and Neptune, most likely reflecting a diminishment in the raw

Figure 3.12   The outer planets — Jupiter, Saturn, Uranus, Neptune — are brought together for contrast. While often lumped together, their sizes and colors clearly separate them into two distinct pairs, J&S very different from U&N. Even within their pairings, they each have unique characteristics, Jupiter its prominent cloud belts and red spot, Saturn its magnificent ring system, Uranus its occasional blandness (tied to a weird axial tilt), Neptune its storm systems and high clouds. Voyager mission/NASA.

material in the ancient circumsolar disk out of which the planets were made.

Though there may be earthlike rocky/icy/metallic cores buried at Jupiter and Saturn's deep centers (which, given the pressure and temperature of 17,000 C for Jupiter, 13,000 for Saturn, would be unlike any rock/ice/metal we could imagine), these planets are — like the Sun — made mostly of light hydrogen and helium. As a result, though Saturn has considerably less mass, it also has less gravity, which allows the light stuff of which it is made to fluff up almost to Jupiter's huge size. Saturn thus has the distinction of having the lowest average density of any major body in the Solar System, Sun included, just 70 percent that of water (Jupiter's and the Sun's twice as great). As teachers love to say, "if you could put Saturn in a bathtub, it would float." In spite of their apparent similarities with the Sun, though, neither can be remotely considered to be a star, as the internal temperatures do not come close to those required to initiate, let alone sustain, nuclear fusion.

What we see from Earth — or with spacecraft for that matter — are the deep gaseous atmospheres, in which hydrogen is in its molecular form, H_2. Neither planet has any solid surface. (What would you weigh on Jupiter? Nothing, as you would just fall into it.) Hundreds of kilometers/miles down, the gases are calculated to compress into liquid form, while farther down yet, roughly a quarter of the way in for Jupiter, maybe halfway for Saturn, molecular hydrogen is expected to turn into a liquid metal whose circulation and rotational motion generate the planets' magnetic fields.

Though J&S both shine to us only by reflected sunlight, they are very much self-luminous if viewed by their natural radio emission or infrared light, the result of internal heat generated by ancient formation, contraction, radioactive decay, and the precipitation of liquid helium. Both are also fast rotators, Jupiter 9 hours 55 minutes, Saturn 10 hours 40 minutes (as measured from the internally-generated magnetic fields). Both spin in the same direction as Earth, Jupiter's axis almost upright (axial tilt 3 degrees), Saturn tilted 27°, which gives Saturn "seasons" that are sort of like Earth's.

Here the two rather part company. Jupiter is covered by colorful bright and dark cloud belts that run parallel to the equator. The rotation period of the equatorial cloud features is a few minutes shorter than the interior's, the result of powerful winds blowing from the west (in the direction of rotation) at up to 350 km/hr (220 mph per hour). The clouds are made of ammonia crystals, while the colors seem to come from a dazzling array of chemicals, including hydrocarbons such as methane, ethane, propane, acetylene, and others. Cloud-top temperatures are very low, −150 C (−240°F) under full sunlight, though they climb rapidly into the interior, which we will never see. Vortices — storm systems — abound. Nothing, though, compares with a huge reddish oval in the southern hemisphere, a spinning high-pressure anti-hurricane twice as big as Earth, the *Great Red Spot*, which has been there for at least 350 years. We also see earthlike lightning on a grand scale, with bolts over 1000 miles long.

Among the most impressive of Jovian features is the planet's magnetic field. Well over a dozen times the strength of Earth's, it holds a magnetosphere of captured solar wind particles tens of times the planet's diameter that would be lethal to an adventurous (or really dumb) astronaut. The field, in part modulated by the inner satellite Io, with which Jupiter is electrically connected, kicks off powerful bursts easily recorded here by radio telescopes. As on Earth, the magnetosphere generates strong polar aurorae.

Orbiting are four large satellites, the "Galilean moons," as well as literally dozens of far smaller ones. In outward order, Io and Europa are as big as our Moon, Ganymede and Callisto of Mercurian size. Taking between two and 17 days to orbit, through the telescope they present an ever-changing and entertaining sight. They are even visible through decent binoculars. Tidally stressed by Jupiter, inner Io is the most active body known, its deadly surface covered with sulfur-laden volcanoes that have given the satellite an orange color. Europa seems to have a deep, warm ocean (life?), while the outer two moons are roughly 50:50 rock and water ice whose surfaces are covered with ancient craters.

Saturn has stronger equatorial winds, 1700 kph (1050 mph) in the direction of rotation, again making the planet look like it is rotating faster than it really is. The cloud belts are muted, in part from a lower temperature (−180 C, −290°F) and a resulting methane haze. The rotation is so fast, and Saturn of such low density, that it is significantly flattened at the poles, the phenomenon noticeable on Jupiter as well. Prominent ovals and giant storm systems are lacking with the distinctive exception of the *Great White Spot*, which develops seasonally during Saturn's 30-year journey around the Sun. Differential winds then spread it around the planet. Unlike Jupiter's, Saturn's magnetic field is fairly weak, less strong than Earth's.

Saturn's glory lies in its extraordinary ring system, which lies exactly in the planet's equatorial plane. Through almost any telescope, the trio of principal rings presents a spectacular sight more than double the size of the planet itself. Fainter rings extend much farther out. The other three big planets also have ring systems that consist of debris kicked off the satellites by stray collisions with other bodies. While these are dark and hard to observe, Saturn's are made of bright, icy rocks typically a few centimeters-to-meters across. They may be

FOUR

SPARKLING STARS

An Entr'acte

When do we first begin to look past the Sun, Moon, planets, into the depths of space to discover the stars? Eight years old, an evening walk with Grandma: I told her that stars had points. She said to look up; see, they don't. With that one response, my life's path within the beauty of the sky had begun. We never know the influences that even the smallest things — actions, remarks — will have. I saw the star again. It was orange. It was magnitude zero. It was Arcturus.

Overlapping the planets in brightness are the generic first magnitude stars. All can be seen from somewhere in the northern hemisphere, all from at least the more southern places in the United States, all from Australia. Northern Americans, Canadians northern Europeans, and those in the British Isles miss quite a few. However, given that first magnitude stars are actually above the horizon, all are visible under any conditions but the worst of smog and city lights. Each has its own story to tell, and together they weave a tale of the history and fate of the Universe, indeed of the Earth and humanity, of our very selves.

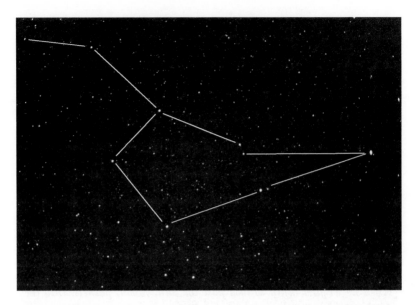

Figure 4.1 Arcturus rising. Looking like a kite, Boötes (the Herdsman) stretches to the left of zero-magnitude orange Arcturus, seen at far right at the convergence of the traditional outline. A classic example of a class K giant, Arcturus is the fourth brightest star of the sky and brightest in the northern hemisphere, just barely topping Vega (Figure 5.3) and Capella (Figure 5.9).

Number

Asking how many "first magnitude stars" (which include those of magnitude zero and the pair at minus one) there are would be kind of like asking for the number of planets. We do not include the Sun. Rather, we've already done that in Chapter 1. While the Sun is indeed a star, the operational phrase is always "Sun and stars." Classically, there are 21 of them, spread across the sky from north to south, from east to west. Another, *Adhara* in *Canis Major*, is right on the border of first and second, which (included here) makes 22. Culturally, the brightest star of second magnitude (*Castor in Gemini*) has last place, in part because of its ancient linkage with nearby *Pollux*. Think of it as being "honorary first magnitude." So now 23. Yet another, *Alpha Centauri*, is easily resolved by a simple telescope into a *pair* of orbiting stars, each of which is of first magnitude. So now we are up to 24. Another double (*Capella*, in *Auriga*) is separable only with sophisticated instruments. Again, both are mag 1, giving us 25. But *Virgo*'s *Spica* is first magnitude only because it is a combination of two second magnitude individuals. One school of thought (if indeed anybody has ever thought of it) might be to toss it out to get us back to 24. But that would be not only silly, but sad, as the star(s) is/are so interesting. If so, Castor should be eliminated too, as it is a sextuple star, with none of the individuals being of first magnitude.

The overall picture is thus clear. As in planetary astronomy, the more you look, the more complex things get. Complexity though, is our main guide to the understanding of Nature; and ultimately it's the guide back to simplicity again. However, before introducing the fabulous 23 (Castor now included, doubles ignored), we pause here

Figure 4.2 Lying on its side, north to the left, Zodiacal Gemini (the Twins) rises. To the upper left are bright Pollux (bottom) and Castor (top). First magnitude Pollux ranks 17th brightest in the sky, while white Castor (23rd) is the brightest star of second magnitude. The rest of the constellation, outlined in Figure 3.4, stretches out in two more or less parallel lines to the right. Castor and Pollux also show up in the auroral picture in Figure 1.6.

to explain the natures of stars and look at their settings, without which the individual descriptions would make little sense.

The Constellations

"You can't tell the players without a scorecard!" Here, the scorecard is the set of fanciful figures, the constellations that organize the stars and lead to their recognition, names, and discovery. Their tale has been told a million times. Now it's 1,000,001.

All societies invented constellations to tell their stories, to express their hopes and dreams, to honor their gods and heroes: the Chinese, Koreans, Indians, Incans, Navaho, each of their sets different from those of the others. Of necessity (not to mention lack of space), we stick here with those of "western" origin, the ones now in place in the (mostly) western hemisphere. As we might expect, their number is not a simple matter. In our current culture, the sky is populated by 88 "official" constellations, some huge, others small, all with nice modern boundaries, all fitting together like a child's cardboard puzzle. Forty-eight come down to us from ancient Greece. Their histories are lost to time, though they likely originated in the far more ancient lands of the middle-east, in Mesopotamia. Beginning roughly with Homer and Hesiod in eighth century BC, the Greeks wove their own cultures into them, the "final" list and number given to us by none other than Ptolemy in his *Syntaxis*.

A common criticism is that most do not look like what they are supposed to be. With some exceptions, they are not meant to portray — they were not, after all, drawn by an artist — but are prominent configurations that represent and symbolize things important to the cultures of the time. Many are special, of them a subset — the *Zodiac* — sacred, as these dozen constellations (introduced in Chapter 1) were believed to be the homes of seven gods, the seven moving bodies of the sky (Moon, Sun, and five bright planets).

Both inside and outside the Zodiac, the traditional constellations provided a basis for late-night stories, the sages telling of how the Great Bear (*Ursa Major*) got into the sky, how winter's *Orion* the Hunter (Figure 5.10) was done in by summer's Scorpius (Figure 5.2), how beautifully Orpheus played summer's *Lyra*, the Harp (Figure 5.3), the list seemingly endless. Among the best is the tale of the rescue of chained Andromeda (the daughter of Queen *Cassiopeia* and King *Cepheus*) by the hero *Perseus*, who on his way back slaying the Medusa (riding *Pegasus*, the Flying Horse), killed *Cetus* (the Sea Monster), who was about to devour her, all as a result of Neptune being annoyed at Cassiopeia for her remarks about Andromeda being more beautiful than the sea nymphs. All are there in the northern autumn sky for your admiration and your own story-telling variations, the figures still alive for our use and enjoyment. Along with crowns, birds, and other real and fantastic animals.

As the Earth turns, the sky seems to go around its own north and south poles that are located above the terrestrial rotation poles (see Figure 1.4). The equators of both the sky and Earth lie halfway between their respective poles. Start in a mid-northern latitude. The angular elevation of the sky's north pole above the northern horizon is the same as the observer's north latitude. From the mid-northern hemisphere, we then see the North Celestial Pole nicely marked by the star *Polaris* in *Ursa Minor*, the Smaller Bear (the star *not* first magnitude, but mid-second). But the South Celestial Pole is the same angle *down* from the southern horizon, so is well out of sight. If a star is close enough to the sky's north pole (closer than the pole is to the horizon), the star just goes under the pole without setting, is always visible, and is said to be *circumpolar* (see Figure 1.4). For most in the heavily populated north, that includes all or parts of both Bears and some of the pieces that make the Andromeda story.

While from a mid-northern location, a lot of the southern sky is visible, there is a similar set of stars around the South Celestial Pole that never rises and is always invisible. If you want to see them all, you travel south to the equator, where both poles are on the horizon and you can admire the whole sky over the course of the day and year. Go farther, into the southern hemisphere, and now it is the South Pole that is up, the northern one down.

Because the Mesopotamians and the Greeks lived well up into the northern hemisphere, they could not see a whole host of stars that circled close around the South Pole. Moreover, there are large patches of the sky in both hemispheres that are accessible to view, but that have few bright stars, and therefore went un-named. If you can't see much, why make a constellation out of it? Besides, the gods needed at least some near-blank canvas onto which to paint anew. The budding scientists of the Renaissance, however, wanted to map *all* the sky. They needed to "fill in the blanks" with new figures about the time that the first deep southern hemisphere explorers were bringing back stories of new constellations that had never before been seen by northerners. Among these *modern constellations* is one of the most famed of all, *Crux*, the Southern Cross (Figure 5.16), holder of not one, but *two* stars of the first magnitude.

The two centuries between 1600 and 1800 were thus an age of invention, in which astronomers of different nations competed to bring glory to the sky, to themselves, and (not surprisingly) to their rulers, sometimes successfully, sometimes not. Dozens, more than a hundred, new figures were created, many appropriate to the times (*Telescopium*, the eponymous Telescope, *Fornax*, the furnace), others nationalistic (*Scutum*, the Shield of the Polish hero John Sobieski), still others just plain silly (how about *Bufo*, the Toad). Space being limited, many were also carved out of ancient patterns. The ultimate addition/ alteration was the replacement of *all* the ancient figures with Biblical characters from both the Old and New Testaments. Blessedly, none survived. To add a bit to the confusion, at least one, *Coma Berenices* (Queen Berenices Hair), while considered modern, is of ancient origin. Ignored by Ptolemy, this lovely lacy cluster of stars just south of the Big Dipper was in later times finally restored to its rightful place.

All was finally put aright in the 1920s by a committee of the *International Astronomical Union* (the folks who de-planeted Pluto), who adopted the 48 ancient figures and cut the modern list to a decent 38. In the meantime, huge mid-southern *Argo*, the Ship of the Argonauts (seen by the Greeks sailing whole across the Mediterranean southeast of Orion) had been divided into the smaller packages of *Carina* (the Hull), *Vela* (the Sails), and *Puppis* (the Stern). So we wind up with our 88, all complete with dotted lines that fence them off and provide homes, now not for the gods, but for all the sky's stars and other wondrous objects.

The ones that got sacked (including Bufo and most of the nationalistic ones) are of course still there for anyone who wishes to seek them out. As are a great number of beloved informal constellations, *asterisms* they are called, such as the *Big* and *Little Dippers* that lie within Ursa Major and Minor.

Star Names

The subject of star names takes entire scholarly books. Here is the nutshell version (rather like the Andromeda myth being told in one sentence). There are three basic kinds of names: proper, Greek letter, and numerical. Proper first. Some are of Greek origin that commonly pertain either to a star's appearance, to its place within its constellation, or to something that makes no sense at all. *Arcturus*, in *Boötes*, the Herdsman (see Figure 4.1), is a fine example. (Meanings are presented in the next chapter.) A few Latinate ones (think *Regulus*, Figure 5.15) sneak in as well. The majority, though are Arabic. The scholars of the great Arab empire that followed Rome had high regard for Greek philosophy and arts, and adopted much of the Grecian constellation lore. They named the stars in part by their locations within the Greek constellations, but to really confuse things, they also named them according to their locations within the Arabs' own indigenous figures. To make matters far worse, the eventual transliterations and translations back into Latin in Renaissance Europe were not, to say kindly, of the best quality. Worse yet, many names were assigned to the wrong stars. It has taken quite some time to regain the original meanings, and much is not known, nor ever will be. All the first magnitude stars (along with hundreds of others) have proper names of some sort, even if they are a bit forced.

Greek letters. Thankfully, the Bavarian astronomer Johannes Bayer had a brilliant organizational idea. He produced a set of gorgeous maps (in his *Uranometria*, published in 1603; see Figure 7.1) on which he plotted the stars observed so carefully by Tycho. In addition he hired a talented woodblock artist to draw in the figures so that we could know where in the sky we were looking, and then used the Greek alphabet coupled with the Latin possessive case of the constellation names. Included were non-Tychonic newly-found stars of the southern hemisphere (which were quite messed up by the initial explorers and Bayer's interpretations). Bayer listed the stars in order of brightness, location, and other things known only to himself. Arcturus, the brightest star in *Boötes* the Herdsman, is also thus *Alpha Boötis* ("Alpha of Boötes"), both names in common use. Others later provided more letters to the new constellations that were being invented.

Numbers. Greek letters give us, with some variations on the theme, names to only a couple dozen or more stars within each constellation. Like a greedy tycoon, we need more, more. After the near destruction of the English Navy from not knowing the position of the fleet, the Crown decided that sailors might need better star positions with which to navigate. In the late seventeenth century, John Flamsteed became among the first to use a precision telescope for the task. Isaac Newton and Edmund Halley (whom we will meet again) "liberated" a preliminary copy of his catalogue and numbered the stars within it from west to east, giving us history's *Flamsteed numbers*. Arcturus is also 16 Boötis. Because Britain is so far north, these numbers go only so far south. Attempts at replicating them in the far southern hemisphere by others failed (largely because of the proliferation of now-defunct constellations), so not all first magnitude stars have them. Beyond these forms are vast numbers of stars named with huge numbers of catalogues. Arcturus has a total of 49 listed names, most of them for specialists only.

A Galaxy of Stars

Another term adopted by Hollywood. Where would the moguls ever be without astronomy! They would of course still be within the *real* Galaxy of stars, which is our local home.

We live in a vast, gravitationally-bound system of stars. It's big, so big that even highly populated with stars it seems at the same time to be almost empty because they are so far apart. All the stars you see, even on a brilliantly clear night in the moonless countryside, belong to it. And these are just the local members. From Chapter 2, the Sun is 8 light minutes away, that is, it takes 8 minutes for light — at a speed of 186,300 miles per *second* (300,000 km/s) — to get to us. The nearest star (first magnitude *Alpha Centauri*, Figure 5.16) is four light-*years* away, nearly 300,000 Astronomical Units, 30 million solar diameters. Space is empty indeed. Most of the ones you see at night are dozens, hundreds, some even more than 1000, light years off. It takes more than 1400 years for the light from first magnitude *Deneb* (in *Cygnus*, the Swan) to get to us.

True, we see the stars as they used to be. Deneb could already have blown up (which it is capable of doing: Chapters 5 and 7) and we might not know about it for more than a millennium. As with the Sun, nobody cares. There is nothing to be done about it anyway. After all, the people around you are seen as of a few trillionths of a second ago, and moreover since they are at different distances, you see them all at different times. Nobody cares here either. If you did, it would drive you bonkers.

Now look at the Big Picture. From side to side, our Galaxy of stars is more than 100,000 light years across. And it really has no edge; it just fades away, making it vastly larger. Within it are 200 to 400 billion stars, our magnificent Sun just one. They come in a vast array of masses, brightnesses, pairings and clusterings, with ages from the just-born to youthful to middle age to dying to the actually dead. We live somewhat over half way out from the center in a quiet neighborhood where stars, typically a few light years apart, not only do not collide, but hardly bother one another.

At first look, stars might appear to be randomly distributed across the sky. They are not. Even some constellations are made of groups of related stars. In the larger picture, most of the two dozen or so first magnitude stars, as well as the majority of the naked-eye stars, are collected along or at least near a broad stream of light that completely circles the sky called the *Milky Way*. No longer visible from town, wiped out by ambient lights, from the dark countryside it can be a stunningly beautiful sight.

Figure 4.3   It's the famed Andromeda Galaxy, but it could just as well be ours. Though two and a half million light years away, "Messier 31" is easily visible to the naked eye. Filled with some 400 billion stars and a vast collection of interstellar gas and dust, it's a flat disk 100,000 light years in diameter set at about a 70-degree angle to the line of sight. Were it ours, the Sun would be somewhat over half way out from the center. The disk is dominated by spiral arms in which new stars are being born, the star-forming regions given away by their blue colors. The arms, outlined by dark dust, are centered on a huge reddish bulge filled with older stars. Two small elliptical (armless) satellite galaxies hover nearby, the lower one M 32. Courtesy of Mark Killion.

From most of the northern hemisphere, the deep southern portion of the Milky Way through the Southern Cross (Figure 5.16) and much of Argo is out of sight, while the part visible in northern winters — through Taurus (Figure 3.1), Auriga, Gemini (Figure 4.2), and east of Orion — is faint and hard to discern. The summertime portion, on the other hand, is striking in its beauty. Falling out of its most northerly reach in Cassiopeia (of the Andromeda myth), it cascades down through Cygnus (the Swan) past Deneb, through *Aquila* (the Eagle with first magnitude *Altair*; Figure 5.4), widening and

Figure 4.4   The Milky Way, the manifestation of the disk of our Galaxy, cascades across the sky from the modern constellation Scutum (the Shield, the most prominent part the right-hand portion of the semi-circle of stars to the left of center) to northern Sagittarius at right. Much of its beauty comes from the ridge of dark interstellar dust that divides it in two and the reddish bright clouds of gas that are illuminated by embedded hot young stars. The one at the lower right corner is seen in extraordinary detail with the Hubble Space Telescope in Figure 4.6.

brightening as it goes through Sagittarius and Scorpius (Figure 5.2). Not the least bit smooth, much of it is broken into two streams separated by an irregular dark band. It and various other dark splotches are so prominent the Incans of South America gave them names, made them "dark constellations."

The Milky Way's mythology is as broad and deep as the stream itself: the pathway of souls to heaven, a celestial river sent by the gods to water the crops. Galileo solved its mystery with his telescope, finding that it was made of countless stars. The Milky Way is the chief manifestation of our Galaxy, so much so that the Galaxy as a whole is commonly called by its name. Its explanation is straightforward. Most — more than 95 percent — of the Galaxy's hundreds of billions of stars are arranged in a flat, thin disk that also contains the off-center Sun. From our perspective, the disk, which has a huge inner bulge, therefore surrounds us in a vast circle. Most of its individual stars are far too faint to see, but their light combines to make a soft illuminated band that encircles our heads. Running down the center of it, dividing it in two, is a thick narrow lane of interstellar gas (made mostly of hydrogen) mixed with mineral dust (initially made by dying stars) that blocks the background starlight. Thick condensations within it, plus more blotches of dust, create the dark clouds, which we now know are stellar nurseries.

Since we are off to the disk's side, 25,000 light years from the center, the Milky Way thus appears quite non-uniform. In the direction away from the center, it's faint and hard to detect (hence its faintness in the northern hemisphere's winter constellations), while in the other direction (toward the central bulge in Sagittarius) it swells with light. From the southern hemisphere, with the bulge and Galactic core overhead, with the Milky Way streaming down both sides like two celestial waterfalls, the sight is spectacular. Only five of our first magnitude stars are found distinctly off its track, while the fainter winter Milky Way is alive with them.

Presented face on, our Galaxy is filled with a set of complex, interlocking spiral arms that seem to wind outward from the center. Our Sun is near the edge of one of them. The whole system is rotating so as to wind up the arms, but over the aeons they keep renewing themselves. Controlled by the inner Galaxy's gravity (the combined gravity of its stars), it takes the Sun (and ourselves) some 250 million years to make a full orbit of the center. Since all the stars around us are on their own peculiar orbits, they move relative to us and to one another. All of our lovely constellations are, over periods of hundreds of thousands, millions, of years thus doomed, destined for breakup. By the time we are on the Galaxy's other side, it's likely that few of the naked eye stars we see now will be visible. And we will have to make up new patterns and star names. Enclosing the whole affair is a thinly populated immense outer halo whose stars go back to the Galaxy's origin some 13-plus billion years ago, the younger disk collapsing from the ancient halo a few billion years later.

More than half a century ago, astronomers detected a strong source of radio radiation coming from the thickest part of the Milky Way in Sagittarius. "Sagittarius A," as it was called in the primitive radio nomenclature of the time, was quickly recognized as the true center of the Galaxy. We now know that the radiation is emerging from the extended surroundings of a central "black hole" whose gravity is so strong that light cannot get out. It's a body with four and a half million Sun's-worth of matter that is smaller than the orbit of Mercury. Such "supermassive" black holes seem to be at the cores of all large galaxies, and may be the organizing principles behind their creation and existence. They may have begun as the spawn of supermassive stars that were formed in the early Universe and that no longer exist. The current crop of massive stars can still collapse into black holes, as seen farther along.

While at this point things seem almost overwhelming, it's just the beginning. Scattered off into the distance, into the millions, even billions, of light years, are more galaxies than there are stars within our own system, nearly a trillion of them accessible with modern telescopes. Four are visible to the naked eye: a pair of small satellite galaxies in the deep southern hemisphere called the *Magellanic Clouds,* the obvious fourth magnitude Andromeda galaxy (see Figure 4.3), and a much fainter one equally close by in the constellation *Triangulum.* All but a few galaxies are moving away from us at speeds proportional to their distances. They take us back in time to the origin of the Universe itself, nearly 14 billion years ago, in the sudden expansion (which goes on today) we now call the Big Bang.

There is far more. Our beloved stars seem to make up less than one percent of the combined energy and mass of the Universe (Einstein's famed equation of Chapter 1 giving the conversion). Another four percent is in the form of hot intergalactic gas. Much of the Universe's mass-energy, though, around 20 percent, is made of a mysterious "dark matter" that radiates no light of any kind but whose gravity controls the structures and internal motions of galaxies. But it all pales beside an even more mysterious "dark energy" that makes the remainder and that appears to be causing the Universe, with its galaxies, to expand ever faster.

Figure 4.5   In this tiny dot of the sky much smaller than the angular size of the full Moon, the Hubble "Ultra Deep Field" discloses 10,000 galaxies that stream off into distances of billions of light years, the nearer and bigger ones rather like ours, each magnificent system reduced to little more than a smudge of light. Though not obvious here, the more distant ones are moving away from us faster than those closer in, the Universe expanding, every galaxy (or group of galaxies) getting ever farther apart. NASA/ESA, S. Beckwith (STScI) and the HUDF Team.

Stars

Personal discovery and admiration of first magnitude stars are enhanced by having an initial outline of stellar natures. We already encountered one star in some detail, the Sun. All stars work like the Sun, and are powered by nuclear fusion. Or once were. Or someday will be. Fusion of light elements to heavier ones at sometime in its life is what makes a star a star.

Through observation and application of explanatory theory, we have learned a great deal, though there is much left to know. Stars are born by gravitational contraction and collapse of the dusty gas clouds that litter the stream of the Milky Way. Young stars are therefore crea-tures of the Galaxy's disk. They are birthed either singly or doubly, or more, or in clusters, most of the latter breaking up and scattering their members into the Galaxy's darkness. As they contract from their birth clouds, as in any compressing gas (think diesel engines) forming stars naturally heat inside until they become hot (several million degrees) and dense enough to fuse their dominant hydrogen (which as in the Sun initially constitutes 90 percent of the atoms) into helium. The new energy source then stops the contraction, and the stars — historically called *dwarfs* — begin their long lives.

And here we arrive at a big rule that explains much of stellar behavior: stars are not just lit by nuclear fusion, they are *supported* by it and by the resulting outward push of internal heat and pressure

Figure 4.6. We look deeply into the Lagoon Nebula (seen as the fuzzy reddish blob toward the lower right hand corner of Figure 4.4), where dark interstellar dust hides the new and forming stars that have condensed out of the associated gaseous cloud. Toward the upper left, a hot star ionizes its surrounding gas, causing it to glow. A. Caulet (ST-ECF, ESA) and NASA.

from fusion that counters the inward pull of gravity. Without fusion, stars must contract under the force of their own gravity until something else halts the fall. Or doesn't.

Masses of new dwarfs range from a mighty top of well over 100 times that of the Sun (the upper limit contended) to a dismal bottom of about eight percent solar, below which they become "substars" that cannot become hot enough inside to run the full nuclear engine. How far down these small critters go, we have little idea, but perhaps as low as our more massive planets. Don't let the name "dwarf" fool you. It's just a term. The most massive dwarf stars radiate power at a rate millions of times that of the Sun. Their surfaces hot and blue in color (violet and blue light having the most energy, red the least), their radii more than a dozen times solar, they may be visible even though thousands of light years away.

Lower the mass, and the luminosity descends rapidly, as does the surface temperature, the dwarf stars' colors going from blue to white to the Sun's yellowish-white. Below a solar mass, they turn yellow-orange, then reddish, with luminosities that can drop as low as less than a thousandth that of the Sun. Under about half a solar mass, *red dwarf* stars put out such low wattages that none is even visible to the naked eye. But Nature loves them. Low mass red dwarfs make the vast bulk of the stellar population, while high mass blue ones are rare. And blessedly so, as they are destined to explode and we don't want to be very close, their rarity reducing the odds. Even solar type dwarfs are not all that common, depending how you define the type, making the order of 10 percent of the whole.

If you stretch out starlight into a long rainbow — a spectrum — you will see it crossed by a "barcode" of dark lines, each individual line uniquely caused by some chemical element or ion (an atom stripped of one or more of its electrons). Measurement of line strength, along with analysis using a healthy dose of atomic theory, allows us to use the lines to find stellar surface temperatures, densities, chemical compositions, rotation speeds, and a variety of other properties.

Astronomers use a century-old alphabetic shorthand that runs OBAFGKMLT to describe the appearance of a spectrum (which lines and ions are present). Class A stars (see Figure 4.7) have powerful hydrogen absorptions, while B stars exhibit hydrogen and neutral helium, O stars ionized helium. Going in the other direction leads to weakening hydrogen and to increased complexity of absorptions of ionized then neutral metals, the sequence ending historically at class M, where molecules dominate.

The classes are firmly linked to the stars' temperatures, which are for the most part solely responsible for the changes in spectral appearance (most chemical compositions being roughly solar). They are always measured on the Kelvin scale. Degrees Kelvin (K) are degrees Celsius above absolute zero of -273 C (-459°F), the system thus avoiding negatives. To get to Celsius from Kelvin, just subtract 273. The classes are also linked to mass and state of ageing. Class O, blue in color and falling between 30,000 and just under 50,000 Kelvin, contains the most massive dwarfs. Blue-white and white B and A dwarfs, whose masses range from a couple to several times that of the Sun, average 20,000 and 8500 Kelvin. The yellow-white Sun, at 5780 Kelvin, lies at the warm end of class G. It is followed by cooler orange-yellow class K (Arcturus and Aldebaran of Figures 3.1 and 4.1 prime examples), then reddish M near 3000–3500 K. At the bottom for real stars is infrared-radiating class L, whose members are either dwarfs or substars. The infrared class T's, so cool as to be invisible to the eye though any telescope, are all substars that are more commonly called "brown dwarfs." They display extraordinarily complex

Figure 4.7 A short piece of zeroth-magnitude Vega's rainbow spectrum stretches out from the invisible ultraviolet at left into blue light at right. Crossing it are a variety of vertical "spectrum lines" that are superimposed by atoms and ions in the outer stellar atmosphere. The strongest are from hydrogen (H). Absorptions of ionized silicon (Si II), Calcium (Ca II), and magnesium (Mg II) are also present. A more detailed look would show hundreds of others. From such a spectrum, we identify Vega as a class A star, and can infer temperature and chemical composition. National Optical Astronomy Observatories, courtesy of E. C. Olson.

molecular spectra that include water and methane. Note particularly that only among dwarfs is class and temperature tied strictly to mass. As stars die, they swell and cool, then shrink, heat, and brighten. So we can also have huge and luminous class K and M "giants" and bigger class M "supergiants" as well as very high temperature low mass stellar cinders.

Now we come to the Great Divide. Look first at dwarf stars under about 10 solar masses, those rather like the Sun. Aside from mass, all the characteristics of other stars are caused mostly by age. Like us, stars live and die. The Sun had enough hydrogen in its core to last for 10 billion years. The radioactive record in primitive rocks, including meteoritic debris from space, tells that we are five billion years old, so we have another five billion left to us. As the hydrogen goes to helium, the core slowly contracts and heats, which contrarily liberates more energy, the star thus gradually brightening and expanding. (Providing humanity does not speed things up, we have about another billion or two years left before the oceans evaporate.)

Let the Sun's internal hydrogen now run out. Over the next couple billion years, gravity makes the old helium-filled core contract, which generates heat that in turn causes the hydrogen fusion to expand into a shell around the dead helium core and makes the outer part of the star expand and brighten to 1000 times its present luminosity. At the same time, the Sun cools at the surface, changing its color from yellow-white to reddish, its class from G to M. As a *red giant*, the Sun grows past the orbit of Mercury, destroying it. But at some point, when the internal temperature hits 100 million degrees, the inert core helium fires up to fuse into carbon and oxygen. The new energy source stabilizes the ageing Sun, causing it to contract a bit, whereupon it quietly consumes its helium as an orange class K giant.

The solar death throes really begin when the core helium runs out. At that point the dead carbon-oxygen core more rapidly contracts and, under the further release of gravitational energy, heats even more. Now surrounded by fusing shells of helium and hydrogen, the Sun becomes thousands of times brighter, reddens further to transform itself into an even bigger red giant that expands to — perhaps even past — the orbit of Earth. And it's goodby to Venus.

Figure 4.8   A swollen future giant Sun, 100 times larger than it is today rises over fields burnt by solar radiation a thousand times more intense than current sunlight. But we'll have to wait more than five billion years to see it. If there is anyone left. From *Stars*, Scientific American Library, New York, Freeman, 1992, copyright J. B. Kaler.

But now we encounter some possible salvation. Stars wind up with much less mass than they start with, compliments of powerful winds that begin with expansion and gianthood and continue vigorously in the final stages of it. At their peaks, the winds can hit a billion times the flow rate of the current solar wind. The future Sun will lose its entire hydrogen envelope, and thus much of its gravitational grip on its remaining planets, causing them to spiral outward. At the same time, it becomes highly unstable and begins to vary in its light.

When the Sun's inward pressures are finally relieved by the huge winds, the nuclear engine shuts down. All that will remain is an ancient carbon and oxygen core just over half the current solar mass. For a time the Sun will be surrounded by rings and shells of ejected matter from aeons of mass loss that will then dissipate into the interstellar gloom. Now called a *white dwarf* no matter what its real color (the term another historical artifact), the ancient still-smoldering core has shrunk to the size of Earth. Supported against further contraction by the outward pressure of its free electrons (those stripped from atoms of carbon and oxygen), an effect caused by the electrons behaving as much like interacting waves as particles, the Sun's only fate is to cool and dim forever. At this point the average density has hit around a million times that of the current Sun (which averages about that of water), giving us a metric ton in a sugar cube. Not rare at all, white dwarfs are everywhere around us. The whole solar affair has taken less than two and a half billion years after the internal hydrogen fuel has run out.

Except for a variety of details, all stars under about 10 (or 8 or 12, the number not well known) solar masses behave this way. The biggest detail is time. More massive stars have larger fuel supplies, but they burn them up so fast that their lives become dramatically shorter. For a star of two solar masses, we are down to a hydrogen-fusing time of only a billion years, at 10 solar, down to 20 million. In the process of dying, many of the more massive giant stars of this set create new chemical elements through extensive networks of nuclear reactions, and launch these, as well as by-products of ordinary fusion, into interstellar space through their winds as fodder for new stars. For a time, the expanding ejecta may glow briefly under the hot remnant's harsh light, thereby producing some of the prettiest of celestial objects. The heavy element content of the Galaxy then increases with time. Without previous generations of stars (especially including high mass ones that explode), those that predate the Sun, our heavy-element planet would not have been possible. At the other end of the scale, below eight or nine tenths of a solar mass, the lifetimes are so long they exceed the age of the Galaxy. None of these has ever died. No wonder there are so many low mass stars.

Really Big Stars

There is, however, a funny thing about white dwarfs. Not surprisingly, as the mass of a star increases, so does that of the nuclear-burning core and, at the end of life, that of the ultimate white dwarf. Double the mass of the Sun, and the final white dwarf comes out at 65

FIVE

BRIGHT STAR

Here, one after the other, are the brightest stars, those of first magnitude, all 23 of them, ranging from Sirius through honorary Castor. How, though, do we organize them? We can go north to south, west to east, by brightness, by temperature, color, any of a variety of things. Position seems best so that we can see them in succession with the turning of the sky. While the celestial vault is continuous, we have to start somewhere, so let's begin in a northern spring late evening with one of the loveliest of stars, wherein we return to the beginning of Chapter 4 and Arcturus (seen in Figure 4.1). From there we will tour more or less to the east in order of rising, or as seen at the same time of night as the year progresses, the stars slowly turning with the seasons (as described in Chapter 1). But we'll also backtrack where we need to, to wherever the winding pathways through the stars may lead, much as the Milky Way meanders across the sky.

We began this adventure by picking a place to stand, and for the most part settled in on the temperate northern hemisphere. But that covers a lot of territory, from the tropics to near the pole. At this point the reader gets to narrow it down. To find a star, it helps to know your latitude, that is, your angular distance from the equator to the pole. You can look it up on a map or find it easily using the Web. The latitude gives the elevation in degrees of the North Celestial Pole above the horizon. Go outside and look to the north, then up by an angle equal to latitude, and if it is dark enough, there is Polaris (see Chapter 4). Moreover, latitude also orients the celestial equator (Chapter 1). The equator's highest arch — where it crosses the meridian to the south — is down from the overhead point by the same number of degrees.

Get the angle that the star makes with the equator from the descriptions of the stars or from the Quick Locator Table at the end of the chapter. It's positive for stars north of the equator, negative if they are to the south of it. Then at 9 PM Standard Time around the predicted crossing date (given approximately within the month as early May, late June, etc.), look north or south of the equator by that same angle. If the equator angle is positive and greater than your latitude, you will have to go through the overhead point and face north. There's the star. (The exact date of 9 PM passage will vary some depending on your east-west position.) For every month past the approximate crossing time, subtract 2 hours from 9 PM, for every month before, add 2 hours.

If the equator angle is less than your latitude minus 90° (considering the minus sign), forget it, the star will not rise above your southern horizon. That is, at 40 degrees north latitude, if the star is south of −50 degrees (40−90) you will never see it. But if the equator angle is positive and greater than 90 minus your latitude (at 40 degrees north, greater than +50), then it never sets to the north and is always visible.

Visual colors, which are an additional aid to identification, are always a source of argument (and are hence argued, sometimes quite vigorously). They are really rather washed out, red more orange-red, orange more yellow-orange, blue more blue-white. It depends some on how high the star is in the sky (passage of starlight through thick air changing the color more towards red), the degree of twinkling, and the sensitivity of the eye, as some people see blue as the sharper color, others red. All this aside, now, go look!

Arcturus (*Alpha Boötis*)

The brightest star of the northern hemisphere (Vega and Capella close behind), ranking fourth in the entire sky, Arcturus shines vividly at zeroth magnitude, –0.04, one of but four stars whose brightnesses go into the negatives (see the star in Figures 4.1 and 5.1). The classic way to locate it is to follow the curve of the Big Dipper's handle to the south. That does not help if you can't see or find the Dipper. But you don't need a signpost, because if Arcturus is up, it's instantly recognizable by its brightness and distinctive orange color. Just 19 degrees above the celestial equator, Arcturus rises in the east in spring evenings as a fine announcer of spring itself. It crosses the meridian at 9 PM (all times Standard) in mid June, and sets in the northwest, passing overhead at the southern tip of the Big Island of Hawaii. The only place it can't be seen is from the innards of Antarctica.

The name comes from the Greek "arktos" for "bear" and "ouros" for "watcher" or "guardian," appropriate for the star that follows Ursa Major, the Great Bear, around the north pole. Consistently, Arcturus is also the Alpha star of Boötes, the Herdsman, who plays the same role as bear driver, the ancient constellation stretching out to the northeast like a giant kite.

The star's brilliance and color come from it being both nearby — only 37 light years away — and an orange class K giant, one of a huge number of such stars. Once thought to be 40 light years distant, Arcturus was used to trigger the opening the 1933 Chicago World's Fair, the star's light having left at the time of the previous fair in 1893. When infrared radiation produced by a fairly cool surface (4300 Kelvin) is taken into account, Arcturus shines with the light of 215 Suns, its radius swollen to 26 times solar. Without too much doubt, our orange friend is quietly fusing its core helium into carbon and oxygen. As an added curiosity, Arcturus's low abundance of heavy elements combined with its motion relative to the Sun suggest that it joined us during a collision with a small galaxy that merged with ours.

Stellar Motions. The term "fixed stars" is a misnomer. While the stars seem to have the same positions they did to the ancient Greeks, with telescopes we can actually see them move relative to the Sun and to one another. Combine their angular motions across the sky with their distances, and we can get their speeds across the line of sight. Subtle shifts in the positions of the barcode spectral absorptions give the speeds along the line of sight (from the Doppler effect, the same thing that makes the pitch of a speeding car go up when it approaches you, drop when it recedes). Combination of the two gives us the actual speed, which then leads us to conclusions about stellar orbits and the rotation and structure of the Galaxy.

Spica (*Alpha Virginis*)

Though we have to backtrack to the west a bit, it's impossible to discuss Arcturus without looking at Spica, the Alpha star of Virgo, the Maiden, one of five first magnitude stars in the Zodiac, six if you count our honorary member, Castor. With a magnitude of 1.04, making it 37 percent as bright as Arcturus, Spica ranks 16th. The two stars, which are physically unrelated, are often spoken of in one breath, as if you continue the curve off the Dipper's handle past Arcturus, you run smack into Spica. Even without the Dipper, recognition is easy, as with Arcturus on the meridian to the south, Spica will be just below and to the right, the star passing 11° south of the highest arc of the equator in late May/early June. The color contrast is among the better ones in the sky, Arcturus a shining orange, Spica a distinctive blue-white. Though a southern star, it's still far enough north to be visible from almost anywhere on Earth. For serious travelers, Spica passes overhead as seen from central Brazil and Peru. Paired with Regulus (Figure 5.15), Spica serves as a locator of the Autumnal Equinox, which lies roughly between the two.

The name, from Latin, means "sheaf of wheat," the star a perfect fit for the constellation. Now holding the Autumnal Equinox, Virgo is a deep fertility symbol that in older times represented the harvest. If your sky is at all a bit dark, look to the southwest of Spica to see the distorted box of stars that make Corvus, the Crow, whose top two stars point back at Spica much as the front two bowl stars of the Big Dipper point to Polaris. Just two degrees south of the ecliptic, Spica is visited monthly by the Moon, which occasionally occults it. Only Leo's Regulus is closer.

Far more luminous than Arcturus, Spica seems fainter mostly because it is so far away, 250 light years, nearly seven times

Figure 5.1 Spica in Virgo. Rising in a horizon glow, Arcturus dominates toward lower left, where some of the constellation's outline (from Figure 4.1) is included for clarity. Spica then appears toward lower right. The clump of stars near top center is the Coma Berenices cluster, while the bright star near the upper right edge is second magnitude Denebola in Leo (see Figure 5.14).

more distant. It's also not one, but a pair of very hot class B stars with temperatures of 22,400 and 18,500 K, making the combination among the hottest and thus bluest stars of first magnitude, beaten out only by Acrux and Mimosa (curiously, both in the Southern Cross: see Figure 5.16). When we take a lot of harsh ultraviolet radiation into account, the hotter of Spica's pair shines with the light of 12 thousand Suns, its mass nearly 11 times solar, the cooler far less in both categories. Both are ordinary hydrogen-fusing dwarfs, though the hotter may be nearing the end.

It takes the Earth a full year to orbit the Sun, Mercury 88 days. The Spica pair takes a mere *four* days, which makes the duo just over a tenth of an Astronomical Unit (AU) apart, a quarter Mercury's distance from the Sun. The Sun is about as perfectly spherical as it can be. Not Spica. The two components are so close that each raises tides in the other, that and high spin rates (for the brighter some 80 times faster than solar) forcing them both into distinct ovals. As the two whip around each other, they present different sized "faces" to the Earth, and thus the system varies in brightness by about three percent, which seems to be enhanced by a slight partial eclipse. On top of that, the brighter is slightly unstable, jittering away with about a one percent variation over a mere four hours. And on top of *that* there may be three more distant, fainter members that can watch the action. When the bigger member of Spica's pair expands as a giant, it will most likely encroach on its lesser neighbor and even lose mass to it, the lesser perhaps becoming the greater. Then when it too ages enough, it'll toss part of itself back to the former big one in a game of celestial pitch and catch. Will it someday produce an exploding white dwarf (see Chapter 4)?

Antares (*Alpha Scorpii*)

Superlatives abounding, one hardly knows where to start. With a magnitude of 0.96, just brighter than Spica, Antares ranks 15th. It's not only the reddest of the half dozen or so first magnitude stars of the Zodiac, it's the most southern of them. Its name is from Greek, "Ant-Ares," meaning "like" or "rival of" Mars (Ares: Chapter 3 and Figure 3.6). The two are easy to confuse, the god of war passing close to it every orbital round. But the star's redness is also perfect as the Heart of the Scorpion, of which it is the Alpha star.

Of the first magnitude stars, Antares is comparable only to magnificent Betelgeuse (see Figure 5.10), the two the only red class M supergiants that can readily be seen. In mythology (so goes one story), Orion was poisoned by the Scorpion, and the gods placed them opposite each other in the sky so that Orion never need look upon his killer. As one star rises, the other more or less sets, and vice versa, so if you see Orion, don't look for Antares. The star is far enough south, 26 degrees below the equator, that even at meridian transit (at 9 PM in mid July), from most of the northern hemisphere it appears low and requires a decent southern horizon. If you have one, the star makes a giant triangle with Arcturus and Spica, Antares at the southeastern apex, its color unmistakable. Crossing overhead as

Figure 5.2 Bright Antares, flanked by the two "Arteries" of the Heart, hovers mid-way up from center, the Scorpion's three-star head dropping more or less vertically at upper right. The rest of the figure falls downward from Antares, then curls around to the left to end at the two-star "Stinger." The blob of stars at lower left is Messier 7, a fine young open cluster (see Chapter 4). Another, Messier 6, is just above it. The field is crowded with the stars of the Milky Way, the path of light broken by dark star-forming dust clouds. Antares is paired with Mars in Figure 3.6.

seen from South Africa and just south of Rio de Janeiro, the star is far enough south so as to be out of sight from a couple degrees south of the Arctic Circle, or from about Fairbanks, Alaska to points north.

We speak of "magnitude" as if it were a singular entity. It is not. The magnitude of a star depends on what you observe it with. Starlight is always a mix of all colors. The human eye is most sensitive to the yellow-green part of the spectrum, so our "visual magnitudes" respond more to the star's yellow-green light than to anything else. After the invention of photography, astronomers began using it to measure magnitudes rather than estimating them by eye. But early photographic emulsions were far more sensitive to a star's BLUE light than to its yellow-green. Cool Antares radiates most of its light in the red and infrared. There is less yellow-green and even less of a blue component to its starlight. In blue light, Antares slips to third magnitude, fainter than its surrounding blue stars. Betelgeuse is similar. In red and infrared light, both become vastly brighter relative to their neighbors. This effect is actually an advantage, as the difference between "blue" magnitudes and "yellow" (visual) magnitudes, as measured by modern devices, provides a measure of perceived color, hence of temperature.

There are more than a dozen magnitude systems that respond to different color light, so many it's hard even for experts to keep track of them all. Then there are absolute magnitudes, the magnitude a star would have at a standard distance of 32.6 light years, bolometric magnitudes that respond to total stellar radiation, and on and on ... (At 32.6 light years, the AU appears to be 1 second of arc across, hence the origin of this weird number.)

If it's dark enough, look for the two blue-white flanking stars immediately to the northwest and southeast, the *Al Niyat*, the "Arteries of the Heart" (Sigma and Tau Scorpii). Farther northwest lies the Scorpion's three-star "head," while to the southeast winds a long curved tail that makes the figure look for all the world like a dreaded scorpion (but you'll have to be south of the Alaskan panhandle to see it all). Antares lies in a rich and wide portion of the Milky Way, which brightens to the east toward Sagittarius (known best by its upside down 5-star *Little Milk Dipper*), with the center of the Galaxy and, quite by coincidence, the Winter Solstice, in between the two. In the other direction, to the northwest, dim Libra, the Scales, weighs in.

"Supergiant" is certainly apt. With a huge mass of 15 to 18 times that of the Sun, well along in its death process following the exhaustion of hydrogen in its core, Antares has swollen to a radius of three or more AU, its diameter at least twice that of its namesake's (Mars's) orbit. It's *so* big that even though 550 light years away, a careful observer can actually detect a disk, in that the star does not wink out suddenly during a lunar occultation (which Antares, not far south of the lunar path, suffers a *lot*). After accounting for infrared radiation from its 3600 Kelvin surface, we find a spectacular luminosity 60,000 times that of the Sun, perhaps (depending on many uncertainties) even greater. Not alone, Antares has a much lesser but still hot and blue companion roughly 500 to 600 AU away from it, the two taking maybe 2500 years to orbit each other. Tough to see with a telescope, contrast effects fool the eye into seeing the companion as green.

It's hard to believe that Antares started life only 10 or so million years ago as a very hot, very blue, class O star. As of now it has grown so much as to become unstable and varies irregularly, enough so as to change its brightness rank by a couple slots. At 15 to 18 solar masses, the star is easily heavy enough to explode as a supernova. We just do not know when. It's most likely fusing its core helium into carbon, but it could be farther along than that. If it's hit silicon fusion, it'll blow in a mere week or two, and then you'll have no trouble finding it at all. Even in the daytime. And Scorpius will be left without a heart.

Vega (*Alpha Lyrae*)

White, pure glorious bright white, magnitude 0.03, fifth brightest star in the sky, Vega ranks number 2 in the northern hemisphere right behind Arcturus (and once thought even ahead of Arcturus, astronomers fooled by the color difference). Rising in the northwest in late spring evenings, 39 degrees north of the celestial equator, the star passes nearly overhead at 9 PM in mid August right through the central US on a rough circle between Washington DC to somewhat north of San Francisco. If you are north of the line, look for the brightest evening summer star you can find just south of the overhead

Figure 5.3   Vega in Lyra, the Harp or Lyre of Orpheus, dominates the picture. This delightful configuration is drawn on the sky by Vega, the star down and to the left of it (the famed "double double star" Epsilon Lyrae), and the parallelogram down and to the right, the pattern obvious enough that the "connect-the-dot" lines are not needed. A fundamental standard, Vega is a mainstay of northern summer skies. See it also in the Summer Triangle in Figure 5.4.

point, and vice versa. The only places it cannot be seen lie from Tierra del Fuego in deep South America to points south.

The name "Vega," in Lyra (the Harp), has nothing to do with a stringed instrument, but derives from the last word of the Arabic phrase "nasr al-waqi," which refers to a "swooping eagle," quite nicely showing how complex star names can be. The star is at the northwestern apex of one of the most renown of large celestial figures, the *Summer Triangle*, which is made of two more white stars, Deneb at the northeastern corner and Altair to the south (see Figure 5.4). In a dark sky, look for Lyra's small, exquisite parallelogram to the southeast of the star, Vega and the ancient harp lying in a rich field at the edge of the Milky Way. Cygnus the Swan, with Deneb and the Northern Cross, flies to the east, while much dimmer Hercules clubs his way across the sky to the west.

Vega, nominally at 9500 K, is the archetypal class A hydrogen-fusing dwarf star, and a magnitude standard against which all other stars are eventually compared. While 230 percent more massive than the Sun, and consequently 36 times more luminous, it's bright mostly because it's close to us, a mere 25 light years away, the sixth closest of the first magnitude stars, right after Fomalhaut.

Vega long presented a puzzle. Its spectrum (see Figure 4.7) showed it to be non-rotating, but too bright for its class. Very careful modern analysis with more sophisticated instruments and theory then revealed that it's actually whipping around with an equatorial speed of 270 kilometers per second, 135 times faster than the Sun, giving it a rotation period of just half a day (compared to the Sun's 25). We were fooled because Vega's rotation axis is pointed almost directly at us. The rotation flattens the star by about 20 percent, and as a result the polar temperature of 10,150 Kelvin is 2200 K higher than the temperature at the equator.

Vega (along with Fomalhaut: Figure 5.6) was one of the first stars found to be surrounded by a warm, dusty disk that radiates infrared light. The Sun is still encircled by such a disk that comes from the time of its formation and that is filled with the debris of disintegrating comets and smashing asteroids (see Chapter 6): and embedded planets. Could Vega have planets too? Fomalhaut has at least one, and nobody would be surprised to see them surrounding Vega either. But don't look for much life, as the star's hydrogen-fusing dwarf lifetime is but 800 million years, far longer than it took complex life forms to develop here.

Altair (*Alpha Aquilae*)

Vega and Altair, Altair the Alpha star of Aquila (the Eagle), are almost two of a kind. Both names, out of Arabic, mean pretty much the same thing, that of Altair coming from the similar phrase for Vega, "al-nasr al-tair," the flying (instead of swooping) eagle. In Altair's case, however, the name is consistent with that of the constellation. Both stars are white, Altair a cooler version of a class A hydrogen-fusing dwarf. And both are nearby, Aquila's luminary just 17 light years away, its

Figure 5.4   The Summer Triangle, seen setting in the west, is made of Deneb at the top, Vega toward lower right, and Altair, at lower left. Altair is instantly recognizable by its two fainter flanking stars, this simple pattern resembling a flying bird that befits Aquila, the Eagle. Deneb sits atop a bigger bird, Cygnus (the Swan), the bright star at its tail. Third magnitude Albireo (a beautiful orange and blue telescopic double) represents the head. Tip it upside down and the pattern becomes the Northern Cross. Lyra's parallelogram (see Figure 5.3) is now to the left of Vega. Toward the right-center edge, find the hand with the pointing finger that makes Delphinus (the Dolphin). Nested within the convergence of the lines approaching Altair flies Sagitta, the Arrow. The Milky Way, with its distinctive rift of dark interstellar dust, runs from upper center toward lower left.

dimmer magnitude of 0.8 and lesser rank of 12th brightest being the result of smaller mass (one and three-quarters solar) and the lower temperature (7550 K) and luminosity that goes along with it, the latter 11 times that of the Sun.

Lying south-southeast of Vega only 9° north of the equator, Altair crosses the meridian to the south at 9 PM in early September, the star renowned as the southern anchor of the Summer Triangle, which also features Vega and Deneb (Figure 5.4). Passing overhead as seen from central America and the northern tip of South America, Altair is visible almost anywhere in the world except in deep central Antarctica.

A somewhat dark sky yields a perfect giveaway, two flanking stars (third magnitude Tarazed and fourth magnitude Alshain, respectively Gamma and Beta Aquilae) that look like the eagle's outspread wings (and that have been taken for wing lights on a high-flying airplane). The remaining part of the constellation spreads to the south across the equator. Nearby are two fainter exquisite constellations, *Delphinus* (the Dolphin, to the northeast), which appears as a hand with a finger pointing south, and *Sagitta* (the Arrow, to the north) that actually looks like what it is supposed to be. In a dark sky, you can see the Milky Way flowing past Altair on its way between Cygnus (with Deneb) and Sagittarius.

Unlike Vega, there is no evidence for any kind of planetary system (however loose that may be in Vega's case), no infrared-radiating cloud. It also falls into a temperature zone in which stars become somewhat unstable, flickering by a tenth of a percent or so with no fewer than nine different oscillation periods all going on at the same time and that range from as short as 50 minutes to as much as 9 hours. But then we come back to some level of brotherhood between the two celestial eagles. Altair has long been known as a fast rotator. Unlike Vega, however, Altair is not rotating with its pole pointed at us, but more "face on" with the axis rather perpendicular to the line of sight, which has a profound effect on its spectrum. An equatorial spin rate of 210 kilometers per second flattens it at the poles and gives it an oblateness of about 14 percent, less than Vega but still substantial (and making temperature determinations and other analyses more than problematic).

Deneb (*Alpha Cygni*)

And now for something completely different. Other than being white, in the Summer Triangle, and being related to yet another bird, Deneb bears little other resemblance to its Triangle-mates Altair and Vega (see Figure 5.4). At the Triangle's northeastern anchor, Deneb — 45 degrees north of the equator — is second only to Capella as the sky's most northerly first magnitude star, the "she-goat" having just one more northerly degree. With a magnitude of 1.25, ranking number 19 (and tied for the honor, such as it is, with the Southern Cross's Mimosa), it is rather far down the list of first magnitude stars, just third from the bottom, making it distinctly number three in the Summer Triangle.

Not simply a member of that large asterism, Deneb is also the top star of two constellations, one formal, the other another well-known asterism, making the star a triple threat. The name, from Arabic, means simply "tail," as befits its position as the tail of Cygnus, the Swan, which with its outstretched wings and long neck really does resemble the graceful avian. But flip Cygnus over, stand it on its head, and you see the beloved informal figure of the *Northern Cross*, with Deneb now at its top. To an approximation, the Northern Cross is opposite its formal-constellation counterpart, the Southern Cross. Deneb's further jewel-like setting within a bright portion of the northern Milky Way only enhances its beauty, as it does that of the Swan itself as the she flies the shining path.

Crossing the meridian at 9 PM in mid-September, Deneb passes overhead through the northern tier of US states on a line from Bangor, Maine, through Minneapolis — St. Paul, to Portland, Oregon, which also takes it across southern Ontario, near Montreal and Ottawa. South of the line look to the north of the overhead point at meridian transit, while north of the line, look to the south. (To confirm its sighting, turn to the west and a bit south to find far brighter Vega.) Deneb is so far north that above the Bangor-Portland line, it's circumpolar, that is, it cannot reach the horizon and never sets. If you have a clear view, look for it scraping the northern horizon around 9 PM in mid-March, when it will be directly below the pole. Even given Deneb's high northerly position, the star is seen everywhere but for the extreme southerly tips of South America and New Zealand, and of course the perpetual Antarctica. From Hawaii and points south into the deep southern hemisphere, both the Northern and Southern Crosses can be seen at the same time.

And what a star! Near 8500 Kelvin, just a bit cooler than Vega, Deneb is among the greatest class A supergiants in the Galaxy. Its ranking of third-brightest in the Summer Triangle is strictly the result of its great (though somewhat uncertain) distance of 1400 light years.

Put where Vega is, it would not only rank first in the Summer Triangle, it would be first among all naked eye stars. Shining at magnitude -8, it would be 15 times brighter than Venus, be as bright as a thin crescent Moon, be easily visible in daytime, and cast modest shadows on a dark night.

The magnificence of the star is the result of its high mass of 15–17 or so times that of the Sun, which gives it a luminosity close to 55,000 times solar and a radius of 110 Suns, half the size of Earth's orbit. While Antares, and to an extent Betelgeuse, are comparable, as class M supergiants they are both so cool that they radiate most of their energy in the invisible infrared (invisible to the eye, but not to appropriate instrumentation). Deneb, however, being white, radiates nearly all its energy in the visible part of the spectrum, where we can see it. As a result, in absolute terms it's far brighter visually than either of the two red supergiants, and among the first magnitude stars ranks as the visually most luminous, by a good margin.

It's hard to know exactly what Deneb is doing at the moment. It might, with a dead helium core, be on its way to becoming a red supergiant, or it might be quietly fusing its helium into carbon. In any case, it's not very old, no more than roughly 10–12 million years of age. There is not much doubt that the star will someday — and probably not too long from now — fuse its core to iron and explode as a supernova, leaving the Swan bereft of its tail.

Fomalhaut (*Alpha Piscis Austrini*)

Three constellations — Capricornus, Aquarius, and Pisces — lay out the "wet quarter" of the Zodiac. But the rains extend farther down. Directly to the south of Aquarius we find the first magnitude star Fomalhaut ("fom-a-lo") as the luminary of Piscis Austrinus, the Southern Fish. No guesswork here, the images on the old maps show the Waterman (Aquarius) pouring the contents of his water jar directly into the fish's mouth, which is the very meaning of the star's name, from the Arabic *fom-al-hut-al-jenubi*, "the mouth of the southern fish." The rest of the constellation faintly spreads 20 degrees or so to the west and requires a dark sky to see. To the south-southwest

Figure 5.5   With a loose imagination, Piscis Austrinus might indeed resemble a fish. Whether it does or not, it holds first magnitude Fomalhaut, the "Fish's Mouth," seen just to the left of center. The string of stars at lower right is part of modern Grus, the Crane, while down and to the left of Fomalhaut is one of the more prominent stars of dim Sculptor (the Sculptor's Studio).

is one of the more obvious of modern constellations, Grus (the Crane), who seems to stalk the southern skies on spindly legs.

Fomalhaut crosses the meridian at 9 PM in late October a full thirty degrees south of the equator. Well off the Milky Way, not surrounded by anything bright at all, the star immediately stands out for its loneliness. Crawling across the southern horizon (for those in mid-northern latitudes), it's a true sign of northern autumn and the chilly days to come. Just a hair more southern than Antares and Adhara, passing overhead as seen from central Argentina and Chile as well as from southern Australia, Fomalhaut can't be seen by anyone north of the Alaskan panhandle and the Aleutians. With a magnitude of 1.16, it ranks 18th, just ahead of Deneb. Another white class A star (with a temperature of 8500 K), it may appear to northerners to have a bit of a false yellowish or orange tint as a result of the passage of its light through a great thickness of air.

Comparison with Vega vividly shows what a difference mass can make. At 25 light years, the two are almost exactly at the same distance, yet Fomalhaut is only a third as bright, the result of dropping the mass from 2.3 solar for Vega to 2.0 for Fomalhaut. With a true spin rate of 100 kilometers per second, Fomalhaut, with a radius 1.8 times solar, takes about a day to make a rotation (double that of Vega). Like Vega, the star was among the first observed to have a surrounding, infrared radiating disk, one with a hole in the middle that led to a speculation of a planet or planetary system.

And sure enough, Fomalhaut has a planet, and who knows, maybe more, making it the third brightest star (just after Pollux and Alpha Cen) known to have one. It's also among the first to show us its planet by direct imaging rather than through a variety of indirect means that involve the gravitational deflection of the star or a transit across its surface. Like so many others, the planet, with the romantic name of *Fomalhaut b*, is of the Jovian variety with a mass under three times that of Jupiter. Unlike Jupiter, it orbits at a huge distance of 115 or so AU from its star, thrice Pluto's average distance from the Sun. As supported by a large and growing amount of data, other planetary systems seem nothing like ours. But then, why should they, given that they all form under a variety of different conditions. We see massive "superearths." Are there real Earths? Suddenly, Fomalhaut does not seem so lonely after all.

Achernar (*Alpha Eridani*)

If you live in New York, Chicago, London, or anywhere in Europe, you can, as they say, forget about it. At 57 degrees south of the equator, Achernar, the Alpha star at the southern end of *Eridanus*, the River, is one of a string of similarly-placed first magnitude stars that include the pair in the Southern Cross, the duo in Centaurus, and brilliant Canopus in Argo, none of which is easily visible anywhere in the US except from southern Texas, Florida, and of course the ever-popular Hawaii. North of these, residents will have to take the southerners' word on it.

And it's too bad, because this hot, blue, class B star ranks ninth in brightness, falling at the lower end of zeroth magnitude (0.46, just

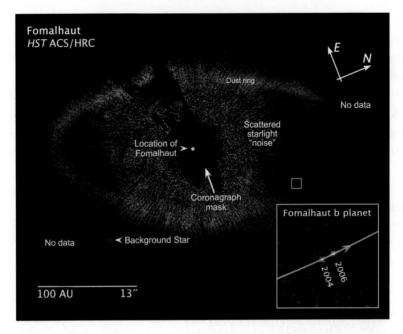

Figure 5.6   With the bright light of Fomalhaut blocked by a masking disk, we see both an encircling dust ring and an actual planet in orbit about the star. NASA, ESA, and Z. Levay (STScI).

brighter than technical first magnitude). The star is *so* far south that its overhead line runs south of the southern tip of South America in the Drake passage between it and Antarctica, and includes no land at all. If you are far enough south, though, Achernar is not hard to locate. From the deep southern US, look for the only bright star near the horizon in mid-evening as November turns to December, way southeast of Fomalhaut.

Too far south to be seen from ancient, or even Alexandrian, Greece (in northern Egypt), the star had no classical name until the first southern explorers saw it. At that time, Eridanus — representing River Ocean — ended at a more northerly third magnitude star called *Acamar* (Bayer's Theta Eridani), whose name derives from an Arabic phrase literally meaning "the end of the River." But to later astronomers, this prominent "new" southern star made a far better, more dramatic, river's terminus, so they extended the stream, transferring the name but spelling it differently to differentiate it from Acamar, which is now just another dock along the way.

Achernar! I was on an observing run at Kitt Peak National Observatory in southern Arizona. And had never seen Achernar. At the Observatory's latitude of almost exactly 32 degrees, the star would for-mally transit the meridian less than a degree north of the horizon. So again, forget it, I'd never see it. BUT! Kitt Peak is 6850 feet (not quite 2100 meters) tall, which allowed me to see nearly one and a half degrees BELOW the formal horizon. And on the horizon, refraction by the Earth's atmosphere lofted it up by nearly another half a degree, giving

Figure 5.7   Looking down from a mountain top to the horizon, with village lights glowing at lower right, Achernar (the end of the River Eridanus at the bottom of the outline) glows fainter than it really is, the result of passage of its light through a great thickness of Earth's atmosphere. The long constellation has its beginning near Rigel (Figure 5.10) in southwestern Orion. The fairly prominent stars of modern Phoenix shine from center to right, the brightest of them second magnitude Ankaa (the Alpha star). The brighter of the pair of stars directly above Achernar, q-1 Eridani, has an orbiting Jupiter-like planet.

an observed elevation for Achernar of almost THREE degrees, plenty enough. Looking at the appropriate transit time, there it was, the end of the River at last, a bit of an effort, but well worth it. You can see the picture in Figure 5.7.

A mid-class B dwarf, Achernar is the hottest of the top nine. The temperature of this hydrogen-fusing dwarf, however, is not well constrained, various estimates falling between 14,500 and 19,000 Kelvin, though the higher end seems the more realistic. The problem once again mostly involves rotation. Spinning at least as fast as 225 kilometers per second, the star is seriously out of round, the directly-measured equatorial diameter of 11.6 times that of the Sun more than half again as large as the polar diameter, the temperature — as for Vega — depending on position. In addition, the rotation has somehow caused the star to eject and maintain a circulating equatorial disk that emits its own radiation, turning Achernar into the brightest of the so-called "Be stars" (for "class B emission"), which can also confuse temperature measures.

Achernar's stellar brilliance despite its distance of 140 light years, and that most of its radiation is emitted in the invisible ultraviolet, is the result of a great luminosity of between 3000 and 5500 Suns. Carrying something like six to eight times the solar mass, not far beneath the limit at which stars explode, it will someday become an immense giant, eject all its outer "unburned" layers (as in Figure 4.9), and die as a massive white dwarf of about a solar mass, one similar to the companion to Sirius but with no companion to itself to watch the action.

Glorious winter. Neglecting for the moment the deep southern stars of Crux and Centaurus, and ignoring the switchback from Arcturus to Spica, we've marched steadily east from the northern spring and summer stars through a couple of autumn's. But now we encounter northern winter and an extraordinary array of bright stars that cascades down the sky from north to south and that rather parallels the winter Milky Way. Though the northern winter Milky Way is much dimmer than the summer version, it's in the winter skies where "first magnitude" most

Figure 5.8   Achernar! In the center of the picture, the star shows its true brightness relative to its surroundings. Comparison with Figure 5.7 then reveals the power of the Earth's absorptive atmosphere near the horizon. The bright star to the lower left is second magnitude Alpha Hydri, which just barely misses being the brightest star in Hydrus, the Water Snake. To the right of Achernar lies Zeta Phoenicis, an eclipsing binary in which two stars in tight orbit mutually get in front of each other every 0.83 days.

heavily concentrates itself, in part because of happenstance, in part because some of the brightest stars (including those fainter than first) are part of a huge physical group. Here we find mighty Orion, one of four constellations with two stars of first magnitude, the Hunter surrounded to the north, east, and south by bright first-magnitude-holding Taurus, Gemini, and the two Dogs, Canis Major and Minor. Add the intense refracting properties of northern winter air, and the stars seem filled with joy to greet you. So given the stars' proximity on the sky, it makes more sense now to change direction, and for awhile go from north to south, from Capella to Canopus, before resuming our easterly trek.

Capella (*Alpha Aurigae*)

Three bright northern stars of magnitude zero divide the sky into crude thirds: Arcturus, Vega, and, at 46° above the equator, the most northerly of them, Capella. Indeed, it is the most northerly of all the bright stars, beating Deneb for the honor by just one degree. At magnitude 0.08, ranking sixth in the sky, this yellow-white class G giant star is also the faintest of the trio, but again not by very much and, given the color difference, really by an unnoticeable amount. From the mid-northern hemisphere, you'll see it rising in late autumn evenings in the far northwest, then crossing the meridian at 9 PM in late January. Like Deneb, Capella passes overhead across the northern US, southern Ontario, and Quebec. And like Deneb, roughly north of the line, Capella is circumpolar, at 9 PM seen dipping under the pole near the northern horizon in late July.

The luminary and Alpha star of Auriga, the Charioteer, Capella's name comes from Rome and means "the She-Goat," Auriga usually depicted as carrying her under his arm. As a fine aid to identification, just to the southwest of the star is a neat longish triangle of fainter stars that represent her three Kids (giving Auriga quite an armful). To the west of Capella lie the star streams that make Perseus, the hero who saved Andromeda from the jaws of Cetus. In a dark sky one can barely detect the Milky Way as it goes from Perseus into Auriga then to neighboring Taurus with reddish Aldebaran to the south-southwest. While Sagittarius, with no first magnitude stars, holds the center of the Galaxy, Auriga (near its border with Taurus) harbors the Galactic Anticenter, where we look outward into the dimmer parts of the Galaxy's disk, thus explaining the Milky Way's relative faintness (which is further suppressed by nearby dark, dusty star-forming regions).

Constellations are in the beholder's eye. There is nothing sacred about them. You can make your own. Just go out and look for patterns and give them names. I did. When first learning them as a child, I discovered what to me is still "the Arrow" (not realizing that Sagitta, the Arrow seen in Figure 5.4, had already been invented). Take Capella in

Figure 5.9   The central portion of Auriga appears as an irregular pentagon with first magnitude Capella at the northwestern apex (the star seen here just up and to the left of center). Up and to the right of Capella bleat the three Kids, arranged in a longish isosceles triangle. The star to the far right in the outlined pattern links Auriga to Taurus. Representing one of the Bull's horns, Beta Tauri (Elnath) is also Gamma Aurigae, the latter name never used. Mars, in Taurus when the picture was taken, hovers near the upper right edge. The smudge just up and to the right of center is the open cluster Messier 38. Set into the faint Milky Way of the Galaxy's Anticenter, Auriga is filled with other clusters and young stars.

Figure 5.9. Draw a line from Capella up and to the right to the left-hand star of the "Kids" (the flat triangle), then to the one at lower right. Finish by using the far left line drawn down from Capella, and another line from the Kids to the bright star near bottom center, the one just below another flat triangle ("the Little Kids?"). And there you have it, an upward-pointing fat arrow that perpetually flies around the pole. And Auriga never will look the same.

There are several reasons for Capella's apparent brightness. First, it's close, just 43 light years away, not that much farther than Arcturus. Second, it's not one star but *two* class G giants so close together that it takes a special effort to see them apart. Third, they are both reasonably luminous in their own rights. Not quite the same, the cooler of the two, Capella Aa (at 4900 Kelvin) is also the more luminous, radiating 93 Sun's-worth of power. But the warmer member, Capella Ab at 5700 Kelvin, is no slacker, giving up about two thirds as much energy, most of it in the visual spectrum.

Double stars. About half of the sky's stars are double — or triple — or more. Castor is sextuple! Doubles and multiples come in a great variety of forms and styles. The components of some, like those of Capella and Acrux (Alpha Crucis), are fairly similar, while others, as in the cases of Sirius and Procyon, have terribly divergent properties. Some doubles are so widely spaced that the individuals take tens, hundreds, thousands, even hundreds of thousands of years to orbit (think Alpha Centauri), while others are in near contact and take just hours. Many are colorful and, like the Antares duo and Cygnus's Albireo, are just plain pretty. And some, like several around Capella called Capella B, C, etc. fool you because they are just in the same line of sight. Doubles are the among the most useful beasts in the stellar zoo, as application of Kepler's rules from Chapter 3 allows the measure of stellar mass, the most important property a star has. We probably owe our lives to their existence, since overflow of mass from a companion onto a massive white dwarf causes the kind of supernova that has created most of the Universe's iron.

The Capella pair's significantly high luminosities in spite of solar-like temperatures are primarily the result of their hefty masses. The two stars are so close to each other, just 0.7 AU apart, about Venus's distance from the Sun, that they are seen to orbit in a mere 104 days, from which (with application of theory) we find their basic properties. Both, with masses of 3.1 and 2.6 times solar, used to be dwarfs on the cool side of class B and not that different from Vega. At that time, born some 500 million years ago, the more massive was the hotter. But their internal core hydrogen ran out, and both have expanded and cooled to become giant stars some 10 times bigger than the Sun with still-dead helium cores.

Since higher mass stars run out of fuel first, the more massive is now farther along toward its death, and is now the cooler of the two.

Expansion slows rotation, though Capella Ab is still spinning fast enough (turning in under 11 days) that like the Sun it is magnetically active and radiates strong X-rays. Aa might be active as well. The two will die as ordinary white dwarfs, unless they expand so much that they encroach on each other and exchange mass or lose it altogether. One may even destroy or absorb the other, leaving one now-massive white dwarf behind. All manner of paths are possible: we and our descendants will just have to wait to find one which will actually be taken.

Aldebaran (*Alpha Tauri*)

Bright (magnitude 0.85, ranking 14th) orange Aldebaran, the luminary of Taurus, the fabled Bull of the Zodiac, is prominent in Figures 3.2 and 3.5. A powerful and enduring mythological figure, Taurus is also barely the westernmost constellation of a ragged circle called the "Winter Six," which includes Auriga, Orion, Gemini, and Canis Major and Minor, each of which has at least one first magnitude star. Colorful, not all that dissimilar from the redness of Mars, Aldebaran crosses the meridian to the south at 9 PM in mid January 17 degrees to the north of the equator, passing overhead as seen from southern Mexico, southern India, and north central Africa. Not quite making it overhead from Hawaii, the star is close enough to the equator to be admired almost anywhere on Earth except from central Antarctica.

But you don't need to look for it "solo," as there are other indicators. Among the most recognizable constellations in the sky, perhaps other than the informal Big Dipper (a part of Ursa Major, the Great Bear), is Orion. Look for his prominent, unmistakable three-star second magnitude Belt that the Arabs called "the String of Pearls." It practically sits on the equator, crossing the meridian to the south at 9 PM in late January. The Belt provides a great "pointer," much like the front bowl stars of the Dipper as they direct you to Polaris. Follow the Belt up and to the right and there is Aldebaran, while down and to the left is the brightest star of the sky, Sirius in Canis Major.

Looking from the other direction, the Pleiades (the Seven Sisters) cluster can be your guide. If the sky is not overwhelmingly bright,

look in early winter mid-evenings for what at first seems like a prominent fuzzy patch of light. A closer examination breaks it up into six
stars, the number most keen-eyed people can see. While Greek
mythology has its "Seven Sisters," daughters of Atlas and mortal
Pleione, their representation in the sky has only six. But there *have* to
be seven, so the missing one became the "lost Pleiad." Binoculars
reveal dozens, telescopes myriads. Then look southeast of it, and
there is Aldebaran. For confirmation, the star is surrounded by yet
another, but much larger cluster, the *Hyades*, mythological half-sisters
to the Pleiades, which form the vee-shaped Bull's head (which in
some representations appears to bear down on Orion, who is about
to club the terrifying animal). While orange Aldebaran makes the
Bull's glaring eye, it is not part of the cluster itself. At a distance of
67 light years, it's in the foreground, the cluster a bit more than twice
as far away. From the dark country, you can see the Milky Way running through the constellation about as faint as it gets, Taurus's
boundary with Auriga closely marking the Galaxy's Anticenter.

A cool (4000 Kelvin) class K giant star 43 solar diameters
across, if placed at the Sun, Aldebaran would extend half way to the
planet Mercury. That would be unhealthy, as the star shines with
the power of 425 Suns, which would fry the Earth quite crisply.
Such might have been the fate of a purported giant planet that was
once tentatively detected through Aldebaran's motion, but one
that has never been confirmed. More relevant, Aldebaran provides
us with a section of the thread of stellar evolution. A star of about
1.7 solar masses, it was once a lot like Altair. But, now some
two billion years old, a hundred million years ago it ran out of its
internal core hydrogen fuel. The contraction of the core generated
more heat, which puffed the outer parts past something that looked
like Capella, then into the true "red giant" (really orange) state it
now finds itself in. In another three million years, after nearly dou-
bling its luminosity and expanding even more (as we might sur-
mise), the star will fire its helium to "burn" to carbon and oxygen,
and then shrink and heat a bit to appear something like Arcturus,
wherein we see that all the stars of the sky are connected in some
way with all the others.

1200 Earth years! Imagine a winter more than 300 of our years long.

Unlike many constellations, and rather like Scorpius, Orion is made of a large super-group of stars more or less related by birth. Betelgeuse seems to have been ejected from a subgroup that lies up and to the right of the Belt, perhaps kicked out by an exploding companion, perhaps through double star interactions. With nearly 20 solar masses, it seems to have no choice but someday to blow itself up, at which point it would shine in our skies as brightly as a gibbous Moon. Antares would be similar. Imagine if they both blew at the same time. On opposing sides of the sky, one would nearly always be visible, leaving us without dark nights.

Rigel (*Beta Orionis*)

Now flip across Orion's Belt a bit to the southwest and across the celestial equator to find the blue supergiant Rigel (really blue-white to the eye), which makes a wonderful contrast with reddish Betelgeuse (see Figures 5.10 and 5.11). Along the way, stop at the Sword, which hangs down from the Belt and is home to the famed Orion Nebula, a vast cloud of illuminated interstellar gas. Though at magnitude zero (0.12) actually a bit brighter than Betelgeuse, Rigel still received the Beta designation from Bayer (either because Betelgeuse was actually the brighter at the time, or because of its southerly position), making it one of five first magnitude stars not called "Alpha." Ranking seventh in brightness, Rigel stands in line only a few percent below Auriga's Capella.

Just eight degrees south of the equator, passing overhead from the southern tropics of central Brazil and Peru, the star is visible from nearly everywhere on Earth except for an uninhabited cap around the north pole in the Arctic Ocean. Being a bit west of Betelgeuse, it comes to 9 PM fruition a bit earlier, in late January, not that it makes much difference, as the Hunter's stars are usually taken together. In mid-temperate zones, at meridian crossing look for it not quite half way up the sky. Not only does Rigel share a constellation with Betelgeuse, it shares a name as well, "Rigel" coming from the Arabic term "rijl al-jauza," the foot of "al Jauza," the Central One.

Figure 5.12   Orion's gem, the Orion Nebula, is a huge cloud of dusty gas several light years across that is ionized (electrified) by energetic ultraviolet radiation from the hot stars in the brightest portion just up and to the left of center, causing it to glow rather like a gigantic neon sign. The region, imaged by Hubble, is a hotbed of star formation, the system seen here less than a million years old. NASA, ESA, M. Robberto (STScI/ESA), and the Hubble Space Telescope Orion Treasury Project Team.

With a spectacular total luminosity of 85,000 Suns (including a lot of ultraviolet light from its 11,500 Kelvin surface, twice the solar temperature), Rigel is one of the four truly magnificent supergiants of the first magnitude family, the others Antares, Betelgeuse, and Deneb, all of which are comparable. The higher the temperature of a body, the more light it radiates. Double the temperature and out comes 16 times more radiation, triple it and it's 81 times as bright. To produce the same luminosity, hot bodies do not have to be as big as cool ones. As the hottest and bluest of the Big Four, Rigel is thus the smallest, its theoretical radius 74 times that of the Sun, 0.34 AU, far smaller than the two red supergiants, and exactly in line with what we measure from the angular size and distance. Though smaller than many "giant stars," Rigel is considered a supergiant largely on the basis of its luminosity and mass, which (to produce that much radiation at that temperature) must be around 18 times that of the Sun, making it one of the heaviest not just of the supergiant quartet, but of all the first magnitude stars. The best guess is that it has a dead hydrogen core and that as a blue supergiant, it is now expanding and cooling at its surface in transition to become a cool red supergiant in the fashion of Betelgeuse and Antares but even grander. With an age of about 10 million years, it does not have a lot of time left to it, however, and like the other massive stars is fated to blow up as a spectacular supernova that would light the landscape as bright as a gibbous Moon.

All this said, astronomy is often an inexact science. Rigel's distance is uncertain by about 10 percent, which means the luminosity is even more uncertain. So are theoretical possibilities. While it's most likely that Rigel is on its way to becoming a red supergiant, it's also possible that it once WAS a red supergiant with a lower initial mass. As observation and theory improve, we will eventually be able to nail it down.

Rigel is not alone, but is actually part of a quadruple system. Not far away (and very hard to see in the glare of the zero-mag star) is a seventh magnitude companion (itself a close double) that is at least 2500 AU away and must take at least 30,000 years to orbit the supergiant. Four times farther away is a vastly fainter low mass star that takes at least a quarter of a million years to make a full turn. Rigel is so bright that it illuminates an arc-like cloud of dust that lies more than two degrees, 40 light years, away to the northwest. Called the Witch-Head Nebula, it is more of a graceful covering that helps warm the foot of Orion.

Sirius (*Alpha Canis Majoris*)

Now what we've been waiting for, Number 1, after Polaris probably the most famed star in the sky, the brightest of them all, "announcing" — SIRIUS. And, given its Alpha status in Canis Major (the Larger Dog), also referred to as the *Dog Star*. Indeed it's the Alpha Star of the entire sky. It's famed from ancient Egypt, as its first sighting in the morning after clearing the Sun foretold the rising of the Nile. So important is Sirius that it has not one, but *two* announcer stars that rise before it and, as if trumpeting for a triumph, tell of its coming over the horizon, so important that one of the announcers is itself first magnitude, Procyon, whose very name means "before the Dog." (The other is second magnitude Mirzam, Beta Canis Majoris.) Even the name tells of its luster, from Greek meaning "searing" or "scorching." Tying the two dogs, Canis Major and Minor, with their master Orion, Sirius is also the southeastern point of the large Winter Triangle seen in Figure 5.11.

At magnitude –1.46, Sirius lies right on the border between minus first and minus second magnitude and is twice as bright as the next star down, Canopus (Alpha Carinae). No matter where you live, if it's up, you can see it, even through the window of a modestly lit room. And only 17 degrees south of the celestial equator, it is visible from nearly everywhere, even from a few degrees north of the Arctic Circle, as it passes overhead in the mid-southern tropics. Go find it southeast of Orion. From the northern hemisphere, his Belt points down and to the left toward it. Not that you need a guide. From a northern clime, look to the south for its 9 PM meridian crossing in mid-February. It's near-impossible to miss.

Because of its brilliance, its low altitude above the horizon as seen from the mid-northern hemisphere, and crackling cold winter skies, Sirius is also the sky's champion twinkler, the star's pure white color broken by refractive dispersion into multiple flashing colors. The effect is strong enough to give the star status as a UFO candidate right up there with Venus, Jupiter, and Mars.

Sirius's brilliance also might make one believe it to be another magnificent star like the earlier Big Four, one perhaps similar to Rigel, which is only slightly bluer. But no. As naked-eye stars go, Sirius is in fact rather modest, actually an ordinary class A dwarf not dissimilar to Vega with a temperature of 9900 Kelvin and a mass of 2.1 solar. The

Figure 5.13   Canis Major, Orion's larger Hunting Dog, seems to stand on his hind legs begging for a treat. Sirius is obvious at top center. To the star's right is one of its announcing stars, Mirzam (Beta), to the left much fainter Muliphein (Gamma). The other announcer, Procyon, appears in Figure 5.11, where with Sirius it makes the Winter Triangle. The faintest of the true first magnitude stars, Adhara, lies at the right-hand (western) apex of the small triangle toward the bottom of the picture. At bottom center is the orange class K supergiant Pi Puppis, in the Stern of Argo, the Ship of the Argonauts. The fuzzy blob down and a bit to the right of Sirius is the open cluster Messier 41.

star's apparent brightness to the eye comes not from any overwhelming luminosity (26 times that of the Sun), but from closeness. A mere 8.6 light years away, Sirius is the fifth closest star system (counting doubles and multiples as one), and — just double the distance of the closest star Alpha Centauri — is the second closest star visible to the naked eye.

Sirius also opens the door to some stellar weirdness. It's a "metallic" star. Unlike Vega, Sirius rotates slowly, which allows some chemical elements to drift downward under the force of gravity, others to rise as their atoms are pushed out by radiation, messing up the true chemical composition. But that pales beside Sirius's motion problem. In the mid-nineteenth century, astronomers noted that Sirius did not travel through space in a straight line, but instead wobbled, evidence that it was being pulled upon and orbited by an unseen companion. Finally in 1862, the best of the nineteenth-century telescope makers, Alvan Clark, spotted eighth magnitude Sirius B, which orbits Sirius every 50 years from an average distance of 20 AU, the little one tough to see in the glare of the Dog Star itself. No one had a problem with it until its temperature could be estimated, which was first found to be similar to Sirius A (and that we now know is a lot more, 25,000 Kelvin). To be dim but hot at the same time requires that it be very small, only about the size of Earth. Tiny and white, Sirius B became the best known of the first three *white dwarfs*, dense, dead stars with fantastic average densities of a million times that of water (see Chapter 4).

Sirius B carries a solar mass, less than half that of Sirius A (which is actually heavy for a white dwarf, such stars limited to under 1.4 solar). Since higher mass stars die first, Sirius B must once have been the more massive of the two. Theory suggests a 6 or 7 solar mass star, one that had overwhelmed our current Sirius A. During its evolution through the giant states, it lost some 80 percent of itself back into space. Perhaps new stars are now being formed using its gift to the cosmos, stellar life coming from stellar death.

Adhara (*Epsilon Canis Majoris*)

Canis Major contains both the brightest of first magnitude, minus first Sirius, and the faintest of the set, Adhara, or Epsilon CMa (seen in Figures 5.13 and 5.14). Most of the generic first magnitude set are the Alpha stars of their constellations. Here and there, where there are two of the "first" in a constellation (as in Centaurus, Crux, and Orion), a Beta pops up. But Epsilon? The fifth letter of the Greek alphabet? Mirzam (Beta) is mid-second magnitude, while Muliphein (Gamma, third letter) is fourth! But both Beta and Gamma are in the northern part of the constellation, along with Sirius, while Adhara is in the south. Clearly Bayer used position here and not so much brightness, as he is usually accused of doing. With an apparent magnitude of 1.50, some would place it in second magnitude, while here we add it to first, our "second magnitude" beginning at 1.51.

Well south, 29 degrees below the celestial equator, the star takes a reasonably clear horizon to see from most of the temperate northern hemisphere, and is completely out of sight for most of Alaska (barring the Aleutians and Panhandle) and for nearly all of Norway and Sweden, the Danes and Britain getting a poor view at best. Adhara passes overhead between the southern tip of South America and Antarctica, right across the Drake strait and around the globe where, again, there is no land at all.

Adhara's luster is sadly neglected in the glare of Sirius. Placed anywhere else but in Canis Major it would be more renowned. Just over 400 light years away, if you put Adhara where Sirius is, it would shine at magnitude -7, some seven times brighter than Venus at her best and be visible in daylight without much having to look for it. Radiating a bluish light from a hot 22,000 Kelvin, Adhara falls among the hottest half dozen first magnitude stars. Adding in its ultraviolet radiation gives it a wonderful luminosity 22,000 times that of the Sun. If you had ultraviolet eyes, Adhara would be the brightest star in the sky. Its simple barcode spectrum, broken by few absorptions, makes it ideal for the study of interstellar gas lying in the line of sight, which superimposes its own very different barcode.

Adhara's mass could be as much as a dozen times that of the Sun. While it is commonly called a "giant," it really is not much of one, and is more what astronomers call a "subgiant": a star that has run out of hydrogen fuel in its core and is preparing to become a real giant in the mode of Betelgeuse or Antares, losing a lot of mass as it goes. While having a lower mass than these two, Adhara is somewhat above the lower edge of the range beyond which stars explode.

There is still some room for it to survive as a massive, rare neon white dwarf (neon the next product beyond carbon and oxygen in a star's nuclear fusion chain). But with an age of about 15 million years, it is just starting to die. We need to wait another couple million to see if we are right about the star's fate.

Higher mass stars are not just uncommon, they are incredibly rare. Nature just does not like to make them. All of the hot blue class O and B dwarfs (those not much hotter than Sirius) constitute only a few tenths of a percent of the dwarf population, and those above the eventual supernova limit (by and large the class O stars) make up far less. Since stars spend most — more than 80 percent — of their lives as hydrogen-fusing dwarfs, high mass evolving and dying stars are even rarer, which explains why there are so few like our next star, Canopus. Those monsters like Deneb, Antares, and Betelgeuse are much rarer. That we see the numbers we do is testimony to their great luminosities: they can simply be seen over vast distances. High mass rarity is a Good Thing. It means that the odds of being close to an exploding supernova are small. The "death zone" is about 30 light years. You would not want to be inside it. The rarest supernovae, those of the highest birth masses, could do damage if at a distance of more than 1000 light years. It has probably happened.

Canopus (*Alpha Carinae*)

Our stack of winter stars started at northern Capella, 46 degrees above the celestial equator, and now stops at Canopus, 53 degrees below the equator and some 100 degrees below Capella. Only between latitudes 37 degrees north and 44 degrees south latitude can you see them both. Canopus is so far south that it is lost for all of Canada, nearly all of Europe, and the northern half of the continental US. But from the southern tier of US states, through Texas,

Figure 5.14 The (northern-hemisphere) winter Milky Way cascades down the picture from upper right to lower left in this picture taken from atop Kitt Peak in southern Arizona, the observatory domes appearing as ghostly shadows. Sirius and Adhara help define Canis Major at upper right. Far below Sirius, Canopus shines near the horizon, the images of the stars somewhat distorted by the wide-angle lens needed to capture them both. Canopus's light is dimmed by the passage of its light through the thickened terrestrial atmosphere near the horizon, as it was for Achernar in Figure 5.7. At center shine the vast number of bright stars of Puppis, the Stern of Argo, while the Vela (the Sails) wafts to the lower left. Canopus is part of Carina, the Hull, whose stars here barely make it above the horizon.

Arizona, and Florida, together with Sirius, which rides above it, the star presents an impressive sight, transiting the meridian at 9 PM in early to mid February. Only at the southern tip of South America is the star seen to fly over any populated land. Canopus also has the distinction of being the farthest star from the ecliptic plane, some 75 degrees south of it and not far from the pole defined by the orbit of the Earth. Next in line is Vega, which is some 60 degrees *north* of the ecliptic. No way can these be confused with the planets of Chapter 3.

Canopus is the luminary of one of the largest of ancient constellations, Argo, the Ship of the Argonauts, and as such was once called (and for that matter could still be called) Alpha Argo. Nothing in the constellation quite comes close to it. Lying southeast of Canis Major, Argo is so very large that in the nineteenth century it was divided into three parts, Carina (the Hull), Puppis (the Stern), and Vela (the Sails). Canopus falls into Carina, and is now properly known as Alpha Carinae.

Far more luminous than Sirius, mid-temperature (7280 Kelvin) Canopus is classed as a rather rare class F "bright giant" that shines with the light of just over 13,000 Suns. With a radius of 73 times that of the Sun, Canopus in our neighborhood would stretch nearly 90 percent to the orbit of Mercury. To achieve the same warmth as Earth gets, our planet would have to be pushed out to more than 110 AU.

Theory shows that the star must carry a mass of around nine times that of the Sun, that it was born around 30 million years ago as a hot blue star similar to the ones that make Orion's Belt, and that it is now dying and in the process of fusing its core helium into carbon and oxygen. It's in an unusual mass range where stars first become red giants like super-versions of Arcturus and Aldebaran, fire their helium, and then shrink, converting themselves back into relatively warm, mid-temperature giants.

Canopus's fate is uncertain. Below birth masses of 8 or 10 solar, stars lose their outer envelopes to reveal their ancient nuclear-burning cores, becoming white dwarfs made mostly of carbon and oxygen. Much above that vague limit, they blow up as supernovae. Canopus, however, is right on the cusp, and it's hard to know which way it will go. There are reasonable odds that the star will at least *try* to develop an iron core, but that it will be stopped after the fusion has converted the carbon and oxygen into a mix of neon and oxygen, Canopus finally expiring as a rare neon white dwarf, having lost 90 percent of itself back into space.

As northern winter starts to turn to spring, we leave the stack of winter stars that started at northerly Capella and dropped south to Canopus, and move north again to begin a new three-star descent in the northern hemisphere, wherein we look at Castor, Pollux, and Procyon, all of which lie close to us. We also for a time leave behind the massive, pre-explosive stars to find a bit of serenity. Plus we encounter a couple oddities that we have not met with before.

Castor (*Alpha Geminorum*) with a Little Help from Pollux

Not first magnitude. But at magnitude 1.58, just a bit fainter than Adhara and ranking 23rd in the sky, Castor is the brightest of second magnitude (see Figure 4.2). But we can't discuss Pollux without Castor, as the two are forever joined as the mythological Twins of the Zodiac. It would be like Abbott without Costello, Fric without Frac, Laurel without Hardy. Not that they are identical twins. In at least one version, Pollux was a god, Castor a demi-god, even human. Astronomically, they could not be much more different either: fraternal twins at best. So we declare Castor an honorary member of the first magnitude clan, rather like Pluto being an honorary planet. And it would be a shame to drop it anyway, as it is such an amazing star, or as we will see, *stars.*

The twins are so close together, not quite five degrees apart (about the separation of the front bowl stars of the Big Dipper), that you can't possibly find one of the two without noting the other, Castor a clear white, Pollux at magnitude 1.14 notably brighter and distinctly orange. Both inhabit the most northerly constellation of the Zodiac, Gemini, which holds the Summer Solstice, the point in the sky where we find the Sun on the first day of northern summer.

We must admit that the Summer Solstice is technically not actually in Gemini. In the latter half of the twentieth century, precession (Chapter 1), which moves the solstices and equinoxes backwards along the ecliptic, took it across the formal border into Taurus. But what do such boundaries mean? Only that the International Astronomical Union adopted an arbitrary line. The Solstice is still closer to the traditional stars of the constellation Gemini than they are to those of the celestial Bull. So for the foreseeable future, the twins have the solstitial ball.

Averaging 30 degrees north of the equator, the pair passes overhead along a line that grazes the northern coast of the Gulf of Mexico and that runs from northern Florida through south central Texas. Circling the globe, the line then skirts the southern shores of the Mediterranean Sea. From north of the line, look in early March to the south of overhead at the usual 9 PM to find them. Later in the night, or in the early evening later in the year, they appear as a pair of celestial eyes, as if the gods are staring back at you. In the Zodiac, even if well northeast of the Solstice, they are visible everywhere on Earth except Antarctica.

Now to Castor itself. At a nearby distance of 51 light years, the "star" we see is not one, but six! Castor is the archetypal sextuple star: no wonder it should be included here. While we focus on the naked eye, pick up a good virtual telescope (or a real one if you can find it), and take a look. Castor then splits into a pair of white class A stars similar to Vega and Sirius, one a bit brighter than other, that are a mere 4.3 seconds of arc (a thousandth of a degree) apart. Long term observation shows that they orbit each other every 445 years at an average separation of 104 AU, two and a half times Pluto's mean distance from the Sun. Kepler's Laws of planetary motion (Chapter 3) applied to the stars gives a combined mass of 5.7 times solar, thirty percent above what we find from luminosity (37 and 13 solar) and temperature (9500 and 8300 Kelvin).

The fit does not seem too good, But wait! Doppler shifts in the bar-code spectrum reveal that each of the two have their own companions, low mass red/orange dwarfs that orbit in mere days and more or less make up the mass difference, giving us decent agreement. But we don't stop there. Off in the distance, a fiftieth of a degree away, is tenth magnitude Castor C, which consists of yet another pair of stars, both low mass red dwarfs that not only orbit each other every 20 hours but actually get in each other's way, each one eclipsing the other every 10 hours or so. This pair, at least 1000 AU away from Castor A and B, then orbits the inner quartet every 12,000 years or more. (We can't tell exactly, as we do not know how the orbit is foreshortened.) Add them up. Six.

In principle it's possible to have more, but a member of a hypothetical septet would have to have an orbit around the inner sextet that would be so big that interactions with other stars would most likely rip it away. You can get more in a multiple star if the stars mix it up all together, but such systems are inherently unstable, and will kick members out through gravitational interactions and cannot live to be very old. So Castor, along with a few others of its kind (notably the Mizar and Alcor pair in the Big Dipper), sets the limit for stable multiples. Not bad for our honorary first magnitude star.

Pollux (*Beta Geminorum*)

Shining at true first magnitude, 1.14, past the middle of the set and ranking 17th, Pollux now seems overwhelmed by Castor. Bayer even gave it second place by calling it Beta Gem after the second letter of the Greek alphabet, probably because they are not all that different and Castor is the more northerly of the two. It's an ordinary orange, quiet, helium-fusing standard class K giant with a temperature of 4800 Kelvin that shines just 45 times more brightly than the Sun, far below the radiance of its seeming classmates Aldebaran and Arcturus. For a giant, it isn't even all that big, just under 10 times the size of the Sun. About all it seems to have going is that, at a distance of just 34 light years, it is the nearest of the class K orange giants, the only reason it makes first magnitude to start with.

So we should at this point just leave Pollux behind and move on to our next star, Procyon. No! Like Fomalhaut and Alpha Centauri B, Pollux has a planet! And it's one of the few giants to make that claim. Like most of hundreds of planets that have been found orbiting other stars, Pollux's was found through minute gravitational Doppler shifts in the star's spectrum caused by the small orbiting body. Pollux then has the distinction of being one of the closest as well as the brightest of planet-holding stars.

Like most of the planets discovered this way, "Pollux b" is a giant Jupiter with a mass at least thrice that of our own Jupiter. Orbiting at a well-separated distance of 1.7 AU, the planet takes 1.6 years to complete a full circuit, making it very different indeed from Fomalhaut's. Are there Earths? Rather *were* there Earth's before Pollux began to grow and brighten as a giant star? Sad to think they may be gone. The same fate will confront our own Earth someday several billion years down the line, but one from which we might, as pointed out earlier, escape.

Procyon (*Alpha Canis Minoris*)

A most interesting star in its own right, Procyon, the luminary of Canis Minor, plays a subservient role to its grander neighbor, Sirius. In the temperate north, rising before Canis Major, as already noted the very name, "pro-kyon" in Greek, means "before the Dog." It's also tied to Sirius through its placement at the northeastern apex of the Winter Triangle, which also includes Orion's Betelgeuse (see Figure 5.11).

Procyon, though, has several distinctions of its own. First, at zeroth magnitude (0.34) it's really quite bright, ranking eighth in the sky in between Rigel and Achernar. Of the first magnitude stars, it's also the closest to the equator. Only 5 degrees north of that celestial divide, Procyon thus has the greatest world-wide visibility of any of the bright stars, excluded from view only within 5 degrees of Antarctica's south pole. From the temperate north, look for it crossing the meridian at 9 PM about halfway up the sky to the south in early March. It does have its own guides. The star lies almost immediately below Castor and Pollux, and the top two stars of Orion, Betelgeuse and second magnitude Bellatrix point leftward at it.

Procyon is another star that is bright mostly because it is nearby. Just 11.5 light years away, it ranks third in distance among the mag-one set. Only Sirius and Alpha Centauri are closer. As naked-eye stars go, it's rather modest. With a temperature of 6500 Kelvin, less than a thousand degrees warmer than the Sun, it radiates at a rate of only seven times solar (as opposed to Sirius's 26 and we won't even mention Deneb, Betelgeuse, and Rigel), with a radius of just double solar and a mass of 1.4 Suns. Adding to its distinctions, rather like Adhara, class F Procyon is on the line between dwarf and "subgiant" in that it seems about ready to give up its core hydrogen fusion. Wait long enough, another 250 million years, and (with distances held constant, which of course can't happen) by the time the star fires its helium into carbon (in its deep core of course), it would reach magnitude -5, becoming the brightest star in the sky.

Notice how all of the first magnitude stars are brighter than the Sun? Most a lot brighter? There is one exception, rather two exceptions, the stars that make Alpha Centauri. It's an example of a "selection effect." Nature shows us what She wants. Stars dimmer than the Sun have to be very close to us even to be visible. But the truly massive, luminous ones can be very far away and still rank brightly. The great rarity of massive stars is more than offset by their luster and their ability to be seen over large distances. The constellation patterns are thus made up almost entirely of stars that dwarf (in the actual sense of the word) the Sun.

The one last distinction, and it is a big one, is that Procyon has a white dwarf companion, one of the first three discovered. With almost supreme oddity, the other one of the trio is the mate to Procyon's doggie counterpart, Sirius. (The third, which was actually the first white dwarf be found, is a companion to fourth magnitude 40 Eridani, in Eridanus, the River.)

At 11th magnitude (14,000 times fainter than Procyon proper), Procyon B is very difficult to observe as a result of its bright neighbor, and for decades remained rather mysterious. Orbiting every 41 years at an average distance of 15 AU, the white dwarf carries a mass of 0.6 solar, much less than does Sirius B, and is just 35 percent bigger than Earth.

Compare Procyon's white dwarf to Sirius's. For ordinary dwarfs (hydrogen burners like the Sun), the bigger the mass, the bigger and brighter the star. White dwarfs show us the reverse. Since there is no nuclear fusion going on to support them, higher masses lead to greater self-gravities that cause a greater squeeze until the free electrons (which act like interfering wavelets: see Chapter 4) can stop the contraction. Higher masses then induce a smaller radius. With less surface area, higher mass white dwarfs of similar temperature are dimmer rather than brighter.

"First magnitude" thus yields a surprise. Some of the brightest stars in the sky have some of the dimmest companions, which in turn helped lead the way to an understanding of stellar life and death. And we are not yet done with them ... read on.

Regulus (*Alpha Leonis*)

We now leave winter's clutch of 10 first magnitude stars, all of which lie close along the plane of the Milky Way, and take a giant leap to the east to visit with lonely Regulus, which sits in solitary splendor in Leo (the Lion) in the Zodiac between the Pollux and Castor pair and Spica. While Canopus and Vega are the farthest from the ecliptic, Regulus is the closest. Only half a degree north of the solar path, the star is regularly passed over by the Moon and visited by the planets. Next in the list of mag-one stars nearest the ecliptic would be Spica (two degrees to the south of the ecliptic), Antares (4.5 degrees south), Aldebaran (5.4 degrees, and just barely occultable given the Moon's orbital tilt), and Pollux (6.5 degrees and lunar-unreachable). Together, Regulus and Spica make a guide to the autumnal equinox, which lies more or less between them. Regulus's name, which has a long and confused history and is one of the few out of Latin, means "the little King," certainly appropriate for its constellation.

Not the brightest of first magnitude bulbs, at magnitude 1.35 Regulus ranks 21st, in between Mimosa and Deneb (which are tied) and Adhara. But it's still first magnitude and the Alpha star of Leo, and its isolation from other bright stars makes it eminently findable. Regulus crosses the meridian at 9 PM in mid-April just 12 degrees north of the equator, passing overhead as seen from central Africa, the northern edge of South America, and Nicaragua. The only place it can't be seen is from central Antarctica.

A class B dwarf a bit on the blue side of white, Regulus's temperature is hard to define. Its spectacular rotation speed of at least 317 kilometers per second (greater than that of any other first magnitude

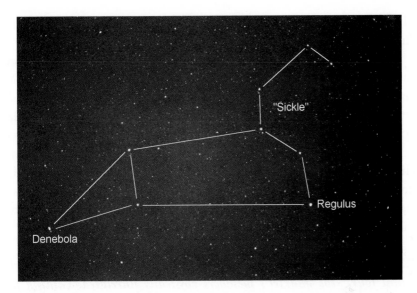

Figure 5.15   Great Leo, the Nemean Lion of the Hercules myth, roars across the picture. At right is the "Sickle of Leo" that forms the Lion's head and that ends in first magnitude Regulus. At the other end is the Beta star, second magnitude Denebola (another "Deneb," meaning "tail," in a different form).

star) bulges and distorts it to the point that the temperature depends severely on stellar latitude, running from 15,400 at the poles to 10,200 at the equator, hotter than the two stars that are at least comparable, Vega and Sirius. Using the temperature to factor in ultraviolet radiation and the distance of 79 light years, Regulus emits the luster of 360 Suns, the radius just under four times solar, the mass coming in at about 3.5 Suns. Though a dwarf with an age of around 250 million years, Regulus seems to be nearing the end of its hydrogen-fusing lifetime.

Not alone, Regulus is a quadruple star with three dim companions. For over 200 years we've know about one of them, an 8th magnitude neighbor nearly three minutes of arc away (4200 AU) that is actually itself double, consisting of two cool dwarfs in a mutual 850-year orbit. The pair must take 120,000 years or more to orbit Regulus proper. The big surprise, found more recently from the stellar spectrum, is that Regulus may also be accompanied by yet another white dwarf, this one very close, about as far as Mercury is from the Sun and taking a mere 40 days to make a lap.

The lightest white dwarfs, which come from the oldest stars to have evolved and died, are calculated to weigh in at around half a solar mass. With a mass estimated at 0.3 times that of the Sun, Regulus's little companion appears to be strongly submassive. A good speculation is that when the white dwarf was a giant, its swollen state allowed it to transfer much of its mass to the star we now know as Regulus. Hitting Regulus on the side, the transferred matter then spun up the star to the fast rotation we see today. When Regulus itself becomes a giant, the flow will be reversed, and the white dwarf will get its matter back with all sorts of attendant activity, some possibly violent.

We now slide far to south to (if you will excuse the sport's analogy) the "final four," the quartet's members physically unrelated. Gliding far beneath Spica and Arcturus, they illuminate just two constellations, the Southern Cross and Centaurus. Like Orion and Canis Major, each of these famed patterns contains two of the first magnitude set, together making a magnificent celestial spectacle. Instantly recognizable, they parade across the southern sky with Acrux (Alpha Crucis) and Mimosa (Beta) in the lead, followed in line by Hadar (Beta Centauri) and Rigil-Kentaurus (better-known as Alpha Centauri). They are so bright that I first saw them through the window of a hotel corridor! Sadly, none of the four is visible anywhere in the United States but for deep southern Florida, and of course Hawaii. Pity the poor yankees, Canadians, and Europeans.

Acrux (*Alpha Crucis*) with a Preview of Mimosa (*Beta Crucis*)

Meet now with that glorious icon of the southern hemisphere, Crux, the Southern Cross, a "modern" constellation that some 400 years ago was taken from the feet of ancient Centaurus, the Centaur. Unlike many of the moderns, Crux stands beautifully on its own, and is so much identified with southern skies that it is displayed on the flags of several nations, notably Australia, New Zealand, and Brazil,

Figure 5.16 Four first magnitude stars appear in one picture, one of such beauty that it should not be marred by constellation outlines. At upper right, the four-star Southern Cross (Crux) shines within the grandeur of the most southerly stretch of the Milky Way. Alpha Crucis, "Acrux," is at the Cross's bottom, while Mimosa, Beta Cru, is up and to the left of it. At left in the picture, the two bright stars are (from right to left) Beta and Alpha Centauri (Hadar and Rigil Kentaurus). Immediately to the left of the Cross is a black splotch in the Milky Way, a complex interstellar star-forming dust cloud popularly called the "Coalsack." It is so prominent that the Incas made a constellation out of it. The funny-looking "star" immediately down and to the left of Mimosa is the compact "Jewel Box" open cluster.

and was a centerpiece in a movement of Richard Rogers' masterpiece score for the television documentary *Victory at Sea*. Set into a bright part of the southerly reaches of the Milky Way amidst star clusters and dark dust clouds, the Cross, with two closely-spaced first magnitude luminaries, presents a stunning sight. Acrux, at magnitude 0.8, ranks 13th, while Mimosa (the Beta star) at magnitude 1.25 falls 19th in the brightness list, tied with Deneb. To these, add second-magnitude Gamma Crucis, which comes in at 24th, right behind Castor.

Acrux, its proper name obviously cobbled from Bayer's Greek letter name, lies a full 63 degrees south of the equator and is the southernmost of all first magnitude stars. It's thus visible only from below a line that lies just north of the Tropic of Cancer, making Hawaii the prime US spot. Crossing the meridian at 9 PM in mid-May, it's a marker of spring in the part of the northern hemisphere where it can be seen (not that in the tropics spring means much) and autumn in the southern. To see it overhead you'd need to walk the coast of Antarctica. Paired to Acrux like Castor is to Pollux, Mimosa has a very similar visibility. It and Acrux are so far south, that both are circumpolar — never setting — from the temperate southern hemisphere.

Acrux's exact brightness (nominally magnitude 0.8, just short of mag zero) is not well known. Indeed, the star presents a bit of a mystery. The problem is that Acrux is not single, or merely double, but triple. Even a small telescope reveals a brilliant blue pair a mere four seconds of arc (0.0011 degree) apart, the brighter at magnitude 1.3, the fainter at 1.8. By itself, the brighter (as the western star, called Alpha-1) makes the first mag list, ranking 20. Maybe. Because the light from each of the stars affects the measure of the other, there is little agreement among observers as to just how relatively bright the stars actually are. The magnitudes given here are just sort of in the middle of the pack. Whatever they are, there is no question about color and heat. Though the temperatures are not well known (near 30,000 Kelvin for Alpha-1, 27,000 for Alpha-2), the pair (both hot, blue, class B) is among the hottest of all the first magnitude stars, giving them their bluish sparkle and causing them to radiate most of their light in the ultraviolet part of the spectrum.

Then the problems multiply. The bar-code spectrum reveals that Alpha-1 is *also* double, but with a very tight orbit, two stars with masses estimated at 14 and 10 times solar spinning around each other every 76 days at in an orbit the size of Earth's. A more or less wild guess tells that the bigger one is pumping out radiation at the rate of 25,000 Suns, the smaller at 7000. Then we have Alpha-2, its luminosity 22,000 Suns (mass about 13), going around the inner pair every 1300 years or more at a distance of at least 400 AU. All seem to be hydrogen-fusing dwarfs or at least close to it. If this isn't enough, the trio may (or may not) be orbited by a slightly cooler four solar mass star 9000 AU away that takes at least 130,000 years to make a circuit. Though steeped in ignorance (ours of course), all of the inner trio at least seem to have the potential to someday blow up as supernovae.

Mimosa (*Beta Crucis*)

Just to the northeast of Acrux lies the Southern Cross's second brilliant member, which rivals Acrux in both luminosity and complexity, though of a different order (see Figure 5.16). Only a bit over three degrees farther north, Mimosa (Beta Crucis) has just about the same visibility as its brighter neighbor, crossing the meridian to the south (as seen from the northern hemisphere) at 9 PM just a few days later, still in mid-May, the Cross's near-twin stars hard to miss. The origin of the name is obscure. Like all deep southern stars that could not be seen from many ancient lands, it carries no traditional proper name. "Mimosa" comes from a Latin word meaning "actor" and somehow seems to derive from a striking tropical flower.

Acrux is the brighter of the two only because it is made of two similar bright stars (plus another). Mimosa stands more on its own, at magnitude 1.25 tied at 19th with Deneb, just beating out Alpha-1 Crucis, which by itself falls 20th. Given no competition from bright neighbors, of the two, Mimosa is the better understood. Still, they are indeed remarkably similar stars, Alpha-1 A (the hotter of the close pair) nearly 30,000 Kelvin, Mimosa standing at around 27,000, the star shining away with the light of 20,000 Suns. Again hot class B, Mimosa rivals Acrux as the bluest of first magnitude (Spica not far behind). With a mass of around 14 Suns, it will almost surely explode. Of the three companions, one is notable, a low mass star about 350 AU distant from Mimosa itself that is in the process of formation. It's so dim, that it cannot be seen with ordinary telescopes in Mimosa's glare, but pops out in the X-ray part of the spectrum where new stars can be quite bright. The evolution of high mass stars is so fast, and that of low mass stars so slow, that Mimosa is already half-way through its hydrogen-fusing lifetime, while its little companion is still getting itself together.

Among the odder aspects of the star is its variability. So far we've encountered four first magnitude stars that change their brightnesses, Antares, Betelgeuse, Altair and one of the components of Spica, the first two varying quite noticeably over periods of years, the last quite subtly. Mimosa is like Spica's member, just more so. Called a "Beta

Cephei star" after the prototype, it varies by tiny fractions invisible to the eye but with at least five periods that range from 4.6 to 8.6 hours. Like a heart gone mad, some parts of the star pulse outward while others at the same time pulse inward.

Hadar (*Beta Centauri*) with Alpha Not Far Behind

Situated between the famed Southern Cross and the nearest star, Alpha Centauri (Rigil Kentaurus), poor Hadar (the Beta star of Centaurus, the Centaur) rather gets lost in the shuffle. It's just usually seen as the precursor to Alpha (as noted by Figure 5.16). And too bad, because it's among the more magnificent of stars. Ranking 10th in the sky, this beautiful hot blue class B star, with a magnitude of 0.61, just fainter than zeroth mag, is the leader of the true first magnitude set.

In the middle of the southern four-pack of first magnitude stars, 60 degrees below the equator, it takes almost nothing to locate as it skims the southern horizon from latitudes of 30 degrees north latitude and a bit south, climbing ever higher with decreasing latitude for its early-June 9 PM crossing of the meridian. Like the others, invisible from farther north, the star flies overhead between the southern tip of South America and the Antarctic peninsula, with only intrepid seafarers to see it.

"Hadar" seems to be some sort of proper name and might even have been identified with another star altogether. If you don't like that one, try "Agena," which refers to the Centaur's ankle. It's rather odd that the two bright stars of Centaurus received more or less proper names, while the two of nearby Crux did not (theirs of quite modern origin). However, all four were visible only from far southern ancient lands (aided by precession, the Earth's 26,000 year wobble, which alters stellar positions relative to us) where the stars were very low above the horizon. Near the horizon, we also must look through a greater thickness of star-dimming air. The Centaurian pair, being the brighter, may thus have made an impact, whereas the dimmer Cruxian stars did not.

Hadar's brightness is indeed intrinsic, but enhanced by it being not one, but three. Right next to the bright star (a mere second of arc away) is a fainter fourth magnitude one that plays little role. However, the barcode spectrum and the *interferometer* (a primitive version of which was used to measure the angular diameter of Betelgeuse using the interfering properties of light) both allow establishment of an orbit that reveals Hadar proper to be made of near twins, each with high temperatures around 25,000 and masses somewhere around 14 times that of the Sun. From a distance of 390 light years, luminosities each come in near 30,000 solar, the stars still hydrogen-fusing "dwarfs": in the technical sense at least. The two whirl around each other every 357 days with separations that change from half an AU to a bit over 5 AU. Once again, we see stars above the critical supernova limit, the one that has a few percent more mass than the other eventually being the first to blow up.

And Finally Rigil Kentaurus (*Alpha Centauri*)

And here we are at the end of this stellar story, the last of our first magnitude stars and the last of the deep southern gang of four (it, Hadar, Acrux, and Mimosa) to slide across the 9 PM meridian, as it does in mid-June, not long after Arcturus. At 61 degrees south of the equator (the second-most southerly of our first magnitude stars), Rigil Kent's visibility to northerners is about the same as the others: don't bother unless you are near or below the Tropic of Cancer. Then, it is more than worth a look at this third brightest star of the sky; at −0.29, it's the brightest of the seven stars of true magnitude zero, and is beaten out only by minus-first Canopus and glorious Sirius. The name comes from an Arabic phrase meaning the "foot of the Centaur," though the star is far better known simply as "Alpha Centauri" or just "Alpha Cen." See it in Figure 5.16.

Yet in a sense we are also at the beginning of the story, as the star (it's renown perhaps topped in the public mind only by Polaris) is the nearest of all and one of the first to have had its distance determined through parallax, the slight angular shift produced by the Earth going around the Sun. It's only 4.32 light years away. "Only." Here is a good chance to put things into cosmic perspective. There are 63,240 Astronomical Units in a light year. Alpha Cen, the "nearest star" lies 275,000 AU away. As of 2012, the Voyager I spacecraft, the most distant man-made object, was 120 AU away. Launched in 1977, there are only 274,880 AU left to go (providing it was launched in the right direction, which it wasn't). No wonder star travel seems so distant, if not impossible.

Yet more perspective. The Sun is just under 0.01 AU wide. Alpha Cen's distance is 30 million times the solar diameter. With such relative separations typical in our part of the Galaxy, it's no wonder that stars simply do not collide, nor even come very close together. And more. There are almost exactly as many inches per mile as there are AU per light year. If you make the Sun an inch (2.5 cm) across, Alpha Cen would be 40 miles (64 km) away, and it would take you half a day of forced marching to walk to the nearest star. Set two one-inch balls 40 miles apart in motion, and try to imagine them colliding. Such numbers also go to the power of the stars. Imagine this time being able to see easily a one-inch ball 40 miles distant.

But again, we are a bit fooled. Alpha Centauri is not only double, but triple. Even a simple telescope sees the bright star break into two, the brighter Alpha Cen A, the fainter Alpha Cen B. Take Alpha Cen B away, and Alpha Cen A, at −0.01, is still zeroth magnitude and the third brightest star, while Alpha Cen B all on its own would be first magnitude (1.33) and the 21st brightest star. Separated on average by 24 AU, they take an easily-detectable 80 years to make a full orbit.

Here, though, we see something very different from the others in the southern quartet, even among all the first magnitude stars. Alpha Cen A is not one of the usual crowd of wonderfully ultra-luminous stars, but instead is a near-solar class G twin just 10 percent more massive than our Sun, while "B" is a somewhat cooler, dimmer, and redder ("oranger?") K dwarf with 90 percent the solar mass. Alpha Cen's brightness is strictly a product of its proximity. Put it a mere 100 light years away, not much farther than Regulus, and it would be invisible to the naked eye.

"Proximity," though, brings up the third star of the trio. Separated from the Alpha A and B pair by 2.2 degrees, about half the angular separation between the front bowl stars of the Big Dipper, lies a dim, 11th magnitude (100 times fainter than the eye can see) star that parallax measures show is about a tenth of a light year — roughly 6000 AU — closer than Rigil-Kentaurus itself. The true nearest star known, it is rightly called "Proxima" Centauri. Studies of its movement show that it is keeping near-exact pace with Alpha proper, and is thus almost certainly a distant, but gravitationally bound, companion.

From its dim luminosity and low 3050 Kelvin temperature, Proxima is a small red dwarf that weighs in at just 12 percent the solar mass. While such stars may seem puny and forgettable, Nature loves them. Seventy percent or more of hydrogen-fusing stars are red dwarfs: yet none — even the brighter ones — is visible to the naked eye. From the system's combined mass and the separation of Proxima, it must take close to a million years to make a full circuit of the bright pair. It's indeed a wonder that it could hold on even after 6 billion years, which again is a testimony to the emptiness of space and the lack of stars that could get close enough to disturb it. Like many of its kind, Proxima Centauri is a "flare star," one capable of magnetically induced flares that massively brighten not just local areas, as on the Sun, but the entire star.

All of which of course brings us back home. While Alpha Cen A, even B, would be amenable to planets with life (very unlike the other first magnitude stars), no earthlike planet in a life-sustaining zone has been found, though "hot earth" in a close 3.2 day orbit is known to orbit Alpha Cen B. And where there is one, there may be another better placed. Were there anyone, Alpha Cen is so close to us that any residents would look outward to constellations that would appear almost exactly like those seen from here. Gazing into the distance, they would also see a brilliant, zeroth magnitude (0.46) star in between the northern figures we call Cassiopeia and Perseus. It's not just any star, but us, our own. What, one wonders, would they name it? Would they wonder too if it harbors life? It of course does, giving us our marvelous platform from which we can admire and appreciate the rest of the Universe.

Quick Locator

Star	Const	Mag	Rank	Color	Equator angle[a]	Meridian Crossing at 9 PM
Achnernar	Eridanus	0.5	9	blue	−57°	late November
Acrux[b]	Crux	0.8	13	blue	−63°	mid May
Adhara[b]	Canis Major	1.5	22	blue	−29°	late February
Aldebaran	Taurus	0.9	14	orange	+17°	mid January
Altair	Aquila	0.8	12	white	+09°	early September
Antares[b]	Scorpius	1.0	15	red	−26°	mid July
Arcturus	Boötes	−0.0	4	orange	+19°	mid June
Betelgeuse[b]	Orion	0.7	11	red	+07°	early February
Capella	Auriga	0.1	6	yellow-white	+46°	late January
Canopus	Carina	−0.7	2	yellow-white	−53°	mid February
Castor	Gemini	1.6	23	white	+32°	early March
Deneb[b]	Cygnus	1.3	19	white	+45°	mid September
Fomalhaut	Piscis Austrinus	1.2	18	white	−30°	late October
Hadar[b]	Centaurus	0.6	10	blue	−60°	early June
Mimosa[b]	Crux	1.3	19	blue	−60°	mid May
Pollux	Gemini	1.1	17	orange	+28°	early March
Procyon	Canis Minor	0.3	8	yellow-white	+5°	late February
Regulus	Leo	1.4	21	blue-white	+12	mid April
Rigel[b]	Orion	0.1	7	blue-white	−08°	late January
Rigil Kentaurus[c]	Centaurus	-0.3	3	yellow	−61°	mid June
Sirius	Canis Major	−1.5	1	white	−17°	mid February
Spica[b]	Virgo	1.0	16	blue-white	−11°	late May
Vega	Lyra	0.0	5	white	+39°	mid August

[a]technically, the *declination*.
[b]Potential supernova.
[c]Better known as Alpha Centauri.

SIX

THE SKY IS FALLING

The vaulted heavens appear so predictable. Stars and constellations wheel around it every night while turning slowly with the year. Come northern spring, there is Arcturus, in winter Betelgeuse, all the stars annually repeating their behaviors. While the Moon and planets move against the starry background, the laws that govern them are immutable. We know where they are going to go, when and where they will appear. And we need no mathematics or calculations; watch them long enough, and their patterns become ingrained into the mind enough to be able to follow them over a lifetime.

Predictability is comforting. But don't we also love surprises? The sky offers them up in abundance, bright things that were not there before and may never be seen again, things that charm, amaze, even frighten. Celestial surprises have two places of origin, the Solar System and the Great Beyond of the Galaxy, both presenting one-time gifts that have the power to make an astronomer of anyone who cares enough to go out to look up, as I first did so long ago.

An evening meteor streaks across the sky, quick make a wish, not that the meteor has any powers of persuasion with the gods. Part of the natural flow of the sky, it — and the more numerous morning meteors — have two sources, asteroids and comets, both of which we can see directly as well. The two share a deep history that goes back four and a half billion years, each giving us a profound connection with our origins.

Failed Planet

That the planets all lie close to the ecliptic, the path of the Sun, immediately implies that they occupy a relatively thin disk that includes the Earth. Otherwise they'd be all over the place, perhaps in the Big Dipper rather than within the Zodiac. Except for retrograde motion caused by the motion of our orbiting Earth, they all also move to the east, all going counterclockwise around the Sun as seen from above the north pole (though at different speeds that depend on distance). The planetary disk thus itself rotates, and in the same direction as the

Figure 6.1   The peaceful evening over the annual Okie-Tex star party is shattered by a brilliant fireball meteor that slashes the sky above Orion's head. It's most likely a small chunk of asteroid no more than meter or so in diameter. Coming in at orbital speed, it shocks and heats a tube of air tens of miles up, the body exploding at the end of its long journey to Earth. Most meteors are from comet flakings and are much fainter, visible only under dark skies. Courtesy of Howard Edin.

Sun, and moreover, close to the Sun's rotational equator. And more still, with three exceptions, the planets all rotate counterclockwise too. These relations tell us that the birth of planets is linked directly to the formation of the Sun four and a half billion years ago, the subject first looked at along with the Moon in Chapter 2.

A homely experiment begins to tell us how and why. Sit in an office chair, put your arms out, and have somebody give you a spin. Bring in your arms, and you spin faster. The effect is called the "conservation of angular momentum," and is the principle behind spins in figure skating and dance and a vast number of other phenomena. Even, it seems, our own existence. Stars form by contraction of blobs (a much-used technical term) found in the thick clouds of dusty gas that lie in the Milky Way (see Chapter 4 and Figure 4.6). As they shrink under the force of their own gravity, they must rotate faster.

Spin a wet bicycle wheel and water flies off at the rim. Spin the Earth, and the stuff at its "rim," or equator, will try to do the same thing. Earth's gravity, however, counters the effect, keeping rock and water from shooting away. But not entirely. In response to its rotation, the solid Earth bulges outward, the equatorial diameter 25 miles (40 kilometers) greater than the polar diameter. Low density Jupiter and Saturn spin so fast (in less than half a day) that you can readily see the effect in a telescope.

Our contracting interstellar blob also bulges and, as the dense core in the middle transforms itself into a real, fire-breathing star, some of it spins out into a dusty circumstellar disk. Unless something disrupts the disk, perhaps the passage of a star in a crowded birthing environment, all forming stars should have one. The dust grains are in orbit, but there are so many (constituting about one percent of the disk's mass) that they collide — and grow. More collisions, more growth. Pretty soon they are of substantial size: a millimeter, a centimeter, a meter. No longer mere dust grains, the bigger ones' increasing gravity attracts lesser particles, accelerating the growth, turning them into *planetesimals*. Those far enough from the Sun, where the disk is cold, attract the abundant hydrogen and helium gases in addition to ices, and grow huge, becoming Jupiter, Saturn, and the rest of the outer gang. There is less raw material out beyond Saturn, so Uranus and Neptune turn out smaller, with more ice and less hydrogen (see Figure 3.12).

Close to the Sun, the boiling heat keeps the gas from being captured, leaving the planetesimals relatively small. After violent collisions clear the competition, the survivors become the rock and metal inner planets. The fierce heat generated in the encounters is enough to partially melt them, sending the heavier iron to the cores, leaving an outer rocky coat on the outside. Similar processes help create the big moons of the outer planets. A final monster collision provides the debris that becomes our own Moon. From theory and our studies of other stars, all of this action takes only about 10 million years. Over the next half billion years, bombardment by the enormous numbers of remaining leftovers creates the craters seen on the Moon, Mercury, Mars, and the outer satellites. And on Earth, too, but erosion and tectonic processes wipe most of them away.

At least that is the standard, parochial version of things. Great numbers of planetary systems around other stars reveal far different orderings for reasons not yet clear. We thought that for the most part "exoplanet systems" would be like ours. So far we have not found "us" out there. More is going on than the simple scenario presented above, including significant planetary migrations out there and here too.

Mercury, Venus, Earth, Mars — all are within 1.5 Astronomical Units. Then Jupiter jumps to 5! Why all this wasted space? What happened to cause it? As they might say in the big bright city, "it ain't empty." A planet might have loved to have formed there, between Mars and the Big One. But Jupiter's immense gravity seems to have kept the rocky raw material stirred up such that the building blocks could only grow so large before collisions induced by the growing giant planet broke them back down again into pieces of interplanetary junk.

As the old year turned to 1801, we began to find them. The first was a rockball at 2.8 AU that was named *Ceres*, the leader of the "minor planets," also called *asteroids*. The biggest of them, all of 580 miles (930 km) wide, Ceres is invisible to the naked eye. So were somewhat smaller Pallas and Juno found in 1802 and 1804, while the next and brightest, Vesta, can be barely seen without aid under perfect conditions. The more we looked, the more we found. We are now at half a million and counting fast as we can. We can see them into the kilometer range, while a vast horde of billions of them range downward to small rocks that lurk undetected.

Jupiter continues to make a mess of the asteroids, mixing them up, continued collisions slowly grinding them into dust. Many have been shown to be no more than rubble piles that have been repeatedly smashed apart only to reassemble under their own weak gravity. Playing the "little planets" like a musical instrument, Jupiter's gravity also resonates within them. There are no permanent asteroids in belts at which their orbital periods would be 1/2, 1/3, 1/4 of Jupiter's, or at average distances of 2.1, 2.5, and 3.3 AU. They can go through these dangerous passages, as well as other more complex ones, but cannot reside there. When one does stray in, thinking it's a fine,

Figure 6.2 All banged up. Asteroid Ida, 58 km (36 miles) long (imaged by the *Galileo* spacecraft as it headed to Jupiter), has suffered mightily through billions of years of collisions. Off to the right is a tiny satellite, Dactyl, a "chip off the old block" that could not escape. Countless other pieces did, some of them perhaps roaming close to Earth: see Figure 1. Galileo Project, JPL, NASA.

empty home, it's immediately kicked out. Sometimes right at us, violent launching by collision doing the same thing.

Arrival

In the textbooks, asteroids are confined to orbits between Mars and Jupiter, most in a "main belt" that stretches between 1.8 and 3.5 AU from the Sun. But in reality, Jupiter scatters them far afield, allowing them to visit not just with Mars, but with Earth as well, a substantial number coming well *inside* Earth's orbit. Just 8000 miles (13,000 km) wide, Earth is a very small target. But Nature has nothing but time, and eventually, one hits. Indeed, many hit.

Asteroids, big ones the size of a house or as small as your fingernail, collide with Earth at orbital velocities in the range of tens of miles per second, tens of thousands of miles per hour, 50 or more times the speed of a supersonic aircraft, fast enough to cross the entire US coast-to-coast in a few minutes. When a body moves through a fluid faster than the natural speed of its wave motion, the fluid's particles can't get out of the way fast enough, and they build up a high-pressure barrier that spreads outward in a *shock wave*. The wake off the prow of a speeding boat is a good example. Another is the sonic boom created by a supersonic jet. It's not produced by the craft "breaking the sound barrier," but is a continuous shock that screams off the front of the fuselage and wings. After the plane passes over, you'll hear the dull "thud" as the shock hits. If the craft is low enough, you'll also hear breaking windows. On a bigger scale, the disturbance in the Earth's magnetic field when hit by a coronal mass ejection from the Sun (Chapter 1) involves similar shock waves.

An asteroidal rock, big or small, hitting Earth's upper atmosphere at Mach 50 some 50 miles (80 km) high sets up an immense shock wave, the energy from which creates a tube of heated and electrified (ionized, with electrons stripped from atoms) air that, added to by the glow of the heated body and its flakings, produces a streak visible on the ground: a *shooting star*, a *falling star*, a *meteor* (from the Greek, a "thing in the air"). Other than the Moon, stars, and planets, meteors are among the most common of celestial sights.

Want to see an asteroid? Stand outside and just watch, especially in the morning hours when you are on the side of the Earth facing in the direction in which we are going around the Sun such that we sweep them up (in the evening they have to catch up with us and seem to go slower). If you are in the deep darkness of the countryside, you'll typically see a few meteors every hour. Most are from comet debris (a subject for later), but about one in 20 might be a visitor from the asteroid belt that is destroying itself.

Most meteors are swift and faint, the parent particles — called *meteoroids* — very small, pea-size down to grains of sand. But on rare occasion, a big one strikes the air, and then it can reach not just first magnitude, but far brighter. The brightness depends on not just size, but what the meteoroids are made of, the velocity relative to Earth, and the angle of entry. The brightest *fireball meteors* are truly spectacular (see Figure 6.1). Streaking across the sky, trailing flakes of heated debris, they will sometimes explode (in part from internal superheated water) with many colors, a natural firework that dwarfs anything produced in a neighborhood celebration. A really big one can light up the ground seemingly as bright as day. The illusion of closeness is powerful. It looks as if the thing might have fallen in your neighbor's yard, whereas in reality, it's maybe 100 miles (160 km) away. These, the ones worth waiting for, are most likely falling asteroids.

I've seen a small number of fireballs. You have to be in the right place under a clear sky and of course be looking more or less up. By the time somebody says, "oh look," the "bolide" is usually gone. My first was as a boy when I saw a burning light falling from the sky that disappeared in a flash of blues and greens. Another time was in the evening seen through the windshield of a car. You just never know when one will appear. There is no way of predicting them. Except for the very largest (the scary ones that hit hard), they are not visible while in orbit around the Sun. Relatively slow, lazy, reddish, their speeds diminished by their having to catch up with Earth, evening fireballs are prettier than the fast white ones of morning that we sweep up. Keep your eye on the sky, and with luck, someday you'll see a really good display.

If you are fortunate enough to witness an outstanding fireball, one that lights the night, wait five minutes, and you may hear the sonic boom hitting the ground. You might even "hear" the passage of the meteoroid itself. As the big rock (or chunk of iron) passes through the Earth's magnetic field, it seems to set up very low frequency radio waves. Broadcast to the ground at the speed of light, they set up magnetic fields and local electrical sparking. For years, scientists thought that these "electrophonic sounds" were illusions, but they are apparently quite real.

Even nighttime is not a necessity. On August 10, 1972, an asteroid estimated to be 10 or more feet across grazed the Earth's atmosphere with such power that it was seen in full daylight as it took under two minutes to go from Utah and back out into space over British Columbia, a distance of 1000 miles (1600 km). There are spectacular pictures of it as it seemed to fly above the Grand Tetons of Colorado. Three other daylight fireballs have been witnessed over the past century, and who knows how many have hit over the oceans that have gone undetected.

On the Ground

If they are big and strong enough, the meteor-producing asteroids (but as we will see, not comet junk) can penetrate all the way through the atmosphere to land on the ground as *meteorites*. Hundreds per day hit Earth, most tiny, most falling in the ocean or on sparsely-occupied land. On rare occasion they hit within sight and if not too heavy, can be picked up off the ground. Slowed by the air and never having had much chance to heat inside, they are cool to the touch. Much more likely, however, meteorites are simply found from long-ago falls. We can then bring one home for admiration, or take it into the lab for close-up study.

What we find is remarkable. The vast majority at first appear as ordinary rocks. A small percentage, however (taking into account the various erosive processes), is made of nearly pure iron with a small admixture of nickel. An even rarer form is a rock-iron combination. From a huge variety of studies, we know that the Earth (and the other terrestrial planets as well) has an iron core about half the size of our planet surrounded by a rocky mantle. Large asteroids apparently developed much the same way, the heat of formation causing their iron to settle to the center. When they violently collide and fracture apart, pieces of rock and free iron are released. Jupiter then helps give them to us.

The stones tell another tale. A good fraction seems never to have been inside a fully-developed asteroid, and has remained fairly primitive. Called *chondrites*, they contain round millimeter-sized inclusions called (not surprisingly) *chondrules* that bear the mark of flash heating and sudden cooling. A small contingent, the *carbonaceous chondrites*, have a high carbon content. The oldest things known, they are dated through looking at the by-products of the decays of uranium and other radioactive elements to be 4.5 billion years old. Since everything seems to have been made at the same time, they give us the age of the Sun and Solar System. Moreover, the weird chondrules — which do not exist in Earth rocks — tell us something of violent processes that accompanied the births of the planets. We just do not know what they might be. Moreover still, primitive meteorites contain minerals from *before* the Solar System was formed, dust grains that give us information on the environment from which the Sun was born. And all from a bright light that once screamed across the sky.

Hard on the Ground
or You Don't Want to be Anywhere Around

How about REALLY big ones? While there are far more small asteroids, an occasional one big as a house plows through. Such was the case in 1908 when a 40,000-tonner hit in the wilds of Siberia. As bright as the Sun they said. Then the great Tunguska meteor exploded in the air, laying waste to the forest for miles around. So violent was the event that it was heard up to 300 miles (500 km) away. But so underpopulated was the area that nobody was killed, though Siberian farmers miles from the impact were apparently knocked down by the shock wave. A smaller version hit in 1947. Lest one think that even bigger events can't occur, look to 0.75-mile-wide Meteor Crater in

Figure 6.3   A piece from a one-time asteroid. Four and a half billion years old, this small rock, filled with oddball round chondrules of unknown origin, is the oldest thing you are ever going to see. Part of the "Allende" fall (named after a small town in Mexico), it's a primitive meteorite that once shared space with Ida and Dactyl in Figure 6.2.

northern Arizona, which was carved out by a tough 60,000-ton chunk of iron that some 50,000 years ago plowed into the solid Earth. Or to the 100-mile filled-in crater off the Yucatan Peninsula of Mexico, which caused such world-wide devastation that 65 million years ago it wiped out the dinosaurs (see Chapter 2 and Figure 2.5).

Will it happen again? Surely. "Potentially hazardous asteroids" of half a kilometer or more flock around us, enough that there are several research groups tracking them down. If that is indeed possible. If we see one on a collision course, destined to hit sometime in the next 10 or 20 years, perhaps we might do something to change its orbit. It would not have be moved much to cause it to miss Earth entirely. However, if it's coming right at us and already at the distance to the Moon ...

Frozen Leftovers

Asteroids represent one of two (possibly overlapping) forms of debris left over from the formation of the planets. Jupiter draws a line of demarcation, a Mason-Dixon line of sorts. Just within its orbit, the heat of the Sun prevented volatile elements and compounds from freezing onto the nascent planetary seeds. From Jupiter on out, however, it was so cold that the developing building blocks could grow fat on them, particularly on abundant water. The moons of Jupiter are half water ice, while the smaller ones of Saturn contain up to 70 percent. Not that there is no water in asteroids, there is, but for the most part nothing like what you find "out there."

Once the outer planets formed, there was still a huge number of small bodies that had not been swept up and incorporated into them. The gravity of the big planets tossed a lot of it out of the planetary system altogether, few of the small pieces ever to be seen again. Beyond Neptune, the disk became so sparsely populated that the planetary seeds could not find many partners to mate with and grow to the size of a real planet. At the same time, gravitational interactions with the leftover debris caused the planets themselves to migrate, Jupiter inward a bit, Saturn and Uranus outward some, and Neptune outward a *lot*, perhaps by as much as 10 AU. Lumbering along like a huge street sweeper, it shoved a lot of the icy debris in front of it, creating a thick belt of the stuff — the *Kuiper Belt* (after the astronomer Gerard Kuiper) — that seems to extend to at least 55 AU from the Sun, beyond which it dramatically thins out. The Belt includes Pluto, which is among the largest of the billion or so pieces that must belong to it, most very small. The planetary interactions naturally explain both Pluto's odd orbit and its orbital locking with Neptune (Chapter 3). A good fraction of the Belt's other members find themselves in the same situation.

Comet Return, Route 1

Collisions among the icy debris as well as gravitational forces shift a few to, and inside, Neptune's orbit, whereupon the planets, playing a small game of pitch-and-catch, move some of the leftovers ever inward. The space between Jupiter and Saturn has a number of pieces of icy stuff in unstable orbits that are moving gradually toward us. Some become tossed into highly elliptical orbits toward the Sun.

Once well inside Jupiter's path, the Sun's heat begins to melt the ices into gases, and the fragile bodies start to fall apart. Energetic sunlight ionizes the gas, stripping electrons from the atoms and chemical compounds to give them electric charges that, along with reflected sunlight on released dust grains, make them glow. And a *comet* is born.

Comet Cocktail

Comets have been known since the first cavefolks looked up into the heavens, and have been recorded, mostly by the Chinese, for more than 2000 years. The word, out of Latin, derives from "hairy star." They do NOT zip across the sky — those are meteors — but drift much like planets against the starry background of constellations. After all, they are interplanetary interlopers with orbits much like that of our own, just more tilted and eccentric.

A typical comet (if there is such a thing, as they are quite individualistic) has a bright and very small central *nucleus* perhaps no bigger than a few miles across. It's surrounded by a fuzzy *coma* of neutral and ionized gas and molecular compounds, much of it water and simple hydrogen. Hit with the solar wind, as well as the solar magnetic field that the wind drags with it, the shining ionized gases are pushed backward, away from the Sun. This *ion tail* can as well precede the comet in orbit as follow it. (Comet tails were our first indication that there *was* such a thing as a solar wind.)

But then there is the solid stuff. A comet is fragile, largely held together by its various chemical ices made of water and other compounds. Destroy this matrix, and the comet releases dust and pieces of rock that get shoved away by gaseous geysers. The pressure of sunlight pushes the dust backward. But the solids have more heft than the gas, and are influenced by the body's orbit. The result is a long, curved *dust tail* that reflects sunlight. All comets thus have not one tail, but *two*, one of gas the other of dust, any dominance of one over the other dependent on the individual cometary makeup. As the comet flies away on its elliptical orbit, it and its tails fade, only to be revived when (or if) the body returns to the Sun.

Figure 6.4   Comet Hale-Bopp, the Great Comet of 1997. Readily visible to the naked eye even in a bright location, HB's two tails (the blue ion tail above, the white dust tail below) span more than a dozen degrees against the background of the constellation Perseus. The blob to the right is the "Double Cluster," a fine sight in a small telescope. More than an Astronomical unit away, the tail stretched a good fraction of an AU in length. For all its glory, the nucleus, the solid cometary body, was only a few tens of kilometers across.

Comets from the Kuiper Belt don't have very big orbits. Their orbital periods around the Sun run from a low of 3 years for Encke's Comet to a couple centuries for those that spend much of their time within the Belt itself. It was in fact the short periods of the comets that stuck to the Zodiac and orbit in the same direction as the planets that led Kuiper and others to figure out that there *was* such a thing as an outer debris belt. Comets lose a significant fraction of their matter every time they get close to the Sun, and thus while numerous, Kuiper Belt comets tend to be small and faint. Quite definitely *not* first magnitude.

But there is another family that not only can be first mag, but can far exceed it. Remember those that got tossed out of the planetary system? They're back.

Comet Return, Route 2

Trillions of ice balls may have been lost to the planetary system. Jupiter and Saturn could have launched them at speeds so great that they left forever for interstellar space, the Sun's gravity not strong enough to bring them back. But Uranus and Neptune, with their smaller gravities, couldn't quite do that. Instead, they created a huge cloud around us (the *Oort Comet Cloud* after Jan Oort, the astronomer who first proposed it) that is filled with perhaps a trillion comet nuclei. Though the cometary nuclei still belong to and orbit the Sun, their Cloud may extend outward for tens of thousands of Astronomical Units, so far that they begin to encroach on the gravitational fields of other stars.

Everything out there is moving relative to the Sun. On occasion, a star — or even a giant cloud of interstellar gas and dust — will come close to the Comet Cloud or even penetrate it. The disturbance can then send a few — or a whole bunch — of comets back into the planetary system on long elliptical orbits. Given the immense orbital sizes, the orbital periods will be very long, tens, even hundreds of thousands of years. Disintegration is long and slow. In addition, many will be back for the first time since leaving nearly five billion years ago. Oort Cloud comets are thus on occasion very bright and impressive, some frighteningly so, enough to give comets an undeserved reputation for bringing doom and destruction. Enough to be first magnitude by far.

Great Comets

"Great Comets" are those you don't have to look for. The tails can stretch longer than an Astronomical Unit and across much of our sky, and the comas and nuclei can far exceed the brightness of Venus. The best known of them is Halley's. Edmund Halley, that friend of Newton who helped liberate Flamsteed's catalogue of stars, was the first to use Newton's laws and the generalization of Kepler's laws to calculate the orbit of a comet. From the accumulated observations, Halley discovered that several, spaced 76 years apart, shared the same path. He concluded that it must be the same one coming back on a closed orbit. His successful prediction of the return in 1758 (sadly after he had died) showed that comets were really a part of the Solar System and forever associated it with his name (most other bright comets named after their finders). Armed with his discovery, Halley's Comet has been traced back for more than 2000 years.

But while at its best amazingly bright and beautiful, while an icon of astronomy itself, among comets it's an iconoclast, not fitting in with either of the two huge comet reservoirs. Its brightness and a backward orbit makes it look as if it came from the Oort Cloud, but its short period of just 76 years makes it look more like a Kuiper Belt comet. It may be a long-period comet from the Oort Cloud that got captured into a shorter orbit by the gravitational action of the Sun's planets. The European *Giotto* spacecraft gave us a close look. For all the ruckus it causes, Halley's is a mere 5×10 mile (8×16 kilometer) wide peanut. Gas fountained from its almost-black crust, while its high-speed dust destroyed the camera. It may someday crack in half. If you are young enough now, maybe you will see it happen!

Halley's is in a class by itself. Its period is roughly the length of a human lifetime, which gives most of us an opportunity to see it once and a favored few the chance to view it twice. Halley's has been recorded throughout much of written history and appears in the Bayeux Tapestry, which chronicles the Norman invasion of Britain in 1066, and in Giotto's "Adoration of the Magi," painted in 1303.

In more modern times, Halley's produced a brand of popular mania. Its 1910 return was one of the showiest ever, as the comet came to perihelion and was brightest when it was close to the Earth. It was used widely in advertising, and there is a story (probably not true) that it inspired a thwarted blood sacrifice of a woman in Oklahoma by a band of fanatics. Comet songs and marches were written and played, and there were cartoons of the evil apparition coming to Earth, knocking down temples, and carrying off small children.

More intriguing, astronomers predicted that on the night of May 18–19, 1910, the Earth would actually pass through the tail. Shortly before, signatures of the deadly gas cyanogen had been discovered in the comet's spectrum, so there was widespread fear that life on Earth might be

Figure 6.5   Halley's dark nucleus, 5 × 10 miles (8 by 16 kilometers) across, sprays out fountains of gas from a solar-heated surface, releasing dust into its magnificent tail. European Space Agency, Giotto Camera Team.

destroyed. On that night some people packed their windows and doorways with wet rags to keep out the fumes, while city sophisticates held rooftop parties to celebrate the end of the world. Entrepreneurs made money selling comet hats and comet pills to ward off the evil effects. Nothing happened. The tail of a comet is so vacuous that no effects were seen at all.

The 1985–86 encounter, one of the most disappointing in its two-millennium recorded history, was seen well only from the southern hemisphere. Yet the hype was enormous and sold a great many telescopes to people who had little idea of how to use them. I had waited nearly all my life for its return, and my first view of it, with binoculars, near the Pleiades star cluster of Taurus, brought tears to my eyes. Hired as a scientific tour guide, from the deserts of Chile I watched it rise tail-first over the Andes mountains framed by the southern Milky Way as it ran like tumbling whitewater through Sagittarius and the Southern Cross. It was so bright that later I could see it from the balcony of a city hotel. Gone now, nearing aphelion, it will return to glory in 2061.

Many other comets have graced the sky only to leave (as far as we are concerned) forever, each intriguingly different. A sampling reveals extraordinary variety. A decade before Halley's return, Comet West of 1976 sported a fantastic fan-shaped dust tail that formed by many sporadic ejections. Rounding the Sun, it broke apart and came back into view in five pieces that will return in half a million years after going out to nearly 14,000 AU from the Sun, a quarter of a light year and a twentieth the distance to Alpha Centauri. (Exact calculation is difficult, as the spewing gases can make the comets into small rockets that unpredictably change their orbits.) From that distance, the Sun would appear not much brighter than Venus does to us.

Five years before that, we admired dusty Comet Bennett, which, in a 1700-year orbit, ran perpendicular to the ecliptic plane. Another five years more into the past, 1965's Ikeya-Seki was among the brightest ever seen, nearly visible to the naked eye in daylight as it tore around perihelion practically grazing the solar surface. It is part of the *Kruetz group*, a fragmented set of a long-ago never-seen giant comet that fell apart and whose progeny still orbit the Sun. Some have even rammed into it. Ikeya-Seki will be back in a millennium.

Twenty years *after* Halley's return, along came Comet Hyakutake of 1996. Rather the reverse of West, it had nothing much of a dust tail, but instead ejected a spectacular gas tail that stretched out for some 40 degrees or more as it plowed along near the Dippers far from the Zodiac. As much as anything, a comet's ion tail is a creature of the Sun's magnetic field, which — dragged out by the solar wind — wraps around the comet like a giant wind sock. But solar fields are in loops that extend outward, then inward. When the comet, going around the Sun, encounters a wall where the solar field switches from one direction to the other, the tail snaps off, disconnects, and flies away while a new one grows in its place. Hyakutake's disconnection events were easily visible to the naked eye. Coming in from the Oort Cloud on a 17,000 year orbit, after receiving a gravitational pounding from the planets, it went back out on a 100,000-year path that will take it 4000 AU away.

Among the champions was Hale-Bopp of the late 1990s which even though an AU away was visible from inside a modestly lit room.

It, and its spectacular combination of gas and dust tails, will make another appearance around the year 4500; see Figure 6.4. But probably *the* great champion of the twentieth century was the Great January Comet of 1910, which indeed *was* visible in daylight. So impressive it was, that it rather overwhelmed the fine appearance of Halley's later in the year, the two commonly confused by the public of the time.

These beauties are more than matched by some that appeared in the nineteenth century, which were certainly enhanced by darker skies. The Great Comet of 1843 had a tail 300 million kilometers — two AU — long, and the central coma of the Great Comet of 1881 was again visible in broad daylight. Certainly one of the more spectacular was de Chéseaux's of 1774, which had six separate dust tails. We never know when another great comet will appear, as the events are truly random, or what it will look like under full solar blast, but this simple survey at least gives some idea of what to watch for when you see the next "first magnitude" comet. Almost certainly one or more will arrive near the Earth in your lifetime. And while it may well be visible from town, it's well worth making a trip to the dark countryside to see and admire it even more.

Collision

Contrary to old belief, comets are benign. They are so small that even if they pass near the Earth, they have no influence on us. Unless of course they hit, which like asteroids, they are quite capable of doing. There are some 200 recognized impact craters on our planet. Unlike those on the Moon, most have been accumulated within the last one or two billion years. Older ones from the great bombardment that took place shortly after the planets formed have been wiped away by erosion: wind, water, glaciation, continental drift. While some are clearly of asteroidal origin, others may well be from comet strikes. Close to home, the Chesapeake Bay structure, a 55-mile-wide buried crater 35 million years old (and tied to another extinction event), is a candidate.

We know from quite direct evidence that such planetary collisions take place. In 1994, a comet took direct aim at Jupiter. Comet Shoemaker-Levy 9 was apparently an ordinary short-period Kuiper Belt comet that got too close to the giant planet, was captured into Jovian orbit, then got tidally pulled apart into at least 22 pieces. Solar gravity stretched the orbit, making it thinner and thinner, until Jupiter just got in the way, and in July of that year we watched the cometary pieces plow one after the other into the planet's cloudy gases. The Solar System's giant bore the dusty scars, some as big as Earth, for weeks before they finally dissipated. More scars from an unseen impact were noted in 2009.

It's widely held that ancient comet impacts (as well as those from "wet" asteroids) brought water to an originally very dry Earth. Without them, there would be no oceans. And the process goes on yet today, albeit at a vastly lower level. Given that long-period Oort Cloud comets are not discovered until they are already pretty close, we won't have as much warning as we do for asteroids. If one does take us on, it will, with no question, be first magnitude!

Figure 6.6 After being captured in orbit around Jupiter, Comet Shoemaker-Levy 9 broke into nearly two dozen pieces. One by one, they rained down onto the cloudy planetary surface, producing long-lasting dusty scars. It could happen here too. Hubble image, H. Weaver (JHU), T. Smith (STScI/NASA).

Comet Flakes

While direct impacts by comets are really very rare, direct "impact" (if you can call it that) by their leavings are most certainly not. If comet nuclei (the ice cores themselves) get close enough to the Sun for us to actually recognize them as comets, with comas and tails, they are doomed. Given enough passages into the inner Solar System, the Sun simply melts them away until they break apart, each piece getting smaller and smaller until it vanishes, as did short-period Comet Biela after its last appearance as a broken two-parter in 1852. When the Earth crossed the comet's path in 1872, Biela was replaced by a spectacular shower of meteors from the debris, for the first time tying meteors and comets together.

But we do not need a breakup to get showered. As comets ply their orbits they leave behind tons of solid particles in the form of dust and small fluffy "rocks" that seem to be little more than compacted grains, the same stuff that makes the dust tail. If and when the Earth gets close to the comet's orbit, it rains its high-velocity supersonic dirt into the atmosphere as meteors. None of the pre-strike "meteoroids" ever reaches the ground as meteorites, only as invisible fine dust that slowly settles downward and can be brought into the lab for study. Since comets are commonly on fast tracks, coming in from the two great reservoirs, the stuff often running perpendicular to the Earth's orbit, meteors from comets are generally faster than are those from asteroids.

The meteoroids that have recently come off comets are on closely parallel tracks that follow the comet's orbit. As they hit Earth, however, perspective makes them appear to come from a *radiant point* in the sky whose position depends on the relative motion of the comet with Earth. Those from Biela's Comet were called the Andromedids, as they appeared to radiate from the constellation Andromeda.

Halley's produces two showers as we pass opposite parts of its orbit, the *Eta Aquarids* that hit around May 5 and the Orionids of late October, which are capable of a meteor every minute or two. By far the best known are the *Perseids* of August 12th or so, which radiate from the constellation Perseus. In a dark sky, they give us up to two meteors a minute from the leavings of Comet Swift-Tuttle, which in a 133-year orbit last came by in 1992. In some cases, like Biela's, the comet is simply gone, with nothing left but the flakes. There are dozens of showers, one always going on. Gradually, the stuff disperses, and early in the morning, when we are on the forward side of Earth, we see lots of so-called sporadic meteors, their parents never to be identified. There are so many cometary meteors that they overwhelm the numbers of those coming from asteroids.

For the most part, meteors from comets are faint, few getting to or exceeding first magnitude. Yet there are some showers that offer up real fireballs. Rather obscure Comet Tempel-Tuttle orbits the Sun and passes Earth's path every 33 years. Dragging behind it are filaments of comet debris. Ordinarily, the comet's shower, the *Leonids* of November 17, is drab if there at all. But every 33 or so years we hit the main junk pile, and unless Jupiter has disturbed the orbit (which it continually does), we will see a spectacular *meteor storm*. During the great storm of 1833, witnesses estimated 100,000 an hour in a celestial snowfall; 1866 was not bad either. Then we missed them for a time until the super-storm of 1966, which was followed by a good show in 1999. The Leonids are known for brilliant fireballs that come from compacted dust-rocks up to a meter in size. It's a long wait for the next display, but like many ephemeral bright celestial objects, more than worth it. As are the awaited objects of the next chapter.

Hot Showers

Here's a brief list of seven well-known meteor showers, identified by the location of the radiant. All are best seen in the early morning. The date is a bit of a rough guide, as it shifts some from one year to the next, and many showers extend over many days. On the other hand, some peak sharply and you need an annual guide, available through various magazines and websites. The "rate" is a typical number per minute, which depends strongly on environment, moonlight, and the annual vagaries of the shower itself. One will occasionally produce a fireball.

Shower	Date	Rate	Parent Comet
Quadrantids[a]	January 4	100	2003 EH1[b]
Lyrids	April 20	10	1861 I
Eta Aquarids	May 5	30	Halley's
Perseids	August 12	80	Swift-Tuttle
Orionids	October 21	20	Halley's
Leonids[c]	November 17	10	Tempel-Tuttle
Geminids	December 13	80	Phaeton[b]

[a]Quadrans is a defunct modern constellation near the Big Dipper.
[b]Dead comet in the asteroid belt.
[c]Capable of thousands per hour every 33 years.

SEVEN

ONCE AND FUTURE STAR

I waited nearly 40 years to see Halley's Comet. I've been waiting 60 to see my first, and probably only, supernova. While it did not reach first magnitude, Supernova 1987a (1987 the year of its eruption) was still an easy naked-eye object. It did, however, have the inconsideration to go off 69 degrees south of the equator, farther down than any of the first magnitude stars, and for me and other temperate northerners quite out of sight. But then we've all been waiting, and far longer, for four centuries, for one to pierce the magic first-magnitude barrier of +1.5, since Kepler's Star of 1604, which was as bright as Jupiter. No other stellar explosion in our Galaxy has since even made it to naked-eye level. I still have some time left ... maybe if I wish upon a star ...

The Heavens, it was once believed, are immutable. Except for the wanderings of the Moon and planets, they are unchangeable. "No, not really," Tycho Brahe (the last of the naked eye observers, whom we met in Chapter 3) might have said in 1572. "Look at this brilliant 'new star' in Cassiopeia just to the north of the Queen's Chair that is as bright as Venus and is visible in the daylight sky."

Whoever "discovered" it, and it surely was not Tycho but probably someone tending to his cows, Tycho had the wherewithal to study it carefully, to follow it, to record its brilliance over its two-year fading time. He did it so very well that in our own era we could reconstruct what he saw. The *nova stellarum*, the "new star," helped pique his interest in astronomy. That led to his great catalogue of

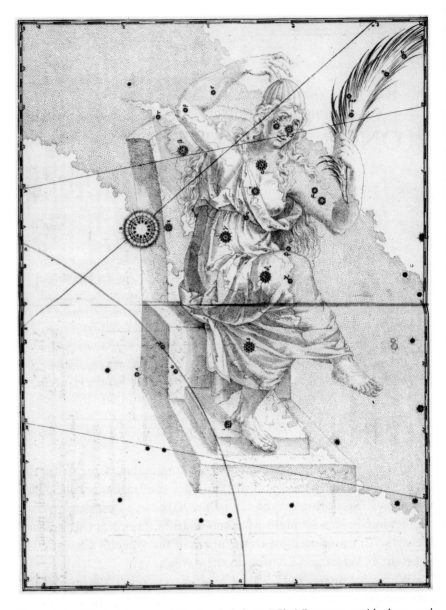

Figure 7.1   The bright stars of Cassiopeia (whose "Chair" appears upside down and does not conform to her drawing) pale beside great Tycho's Star of 1572, which here hardly needs direction to find. Long gone, it is one of the two most recent Galactic supernovae, the other being Kepler's Star of 1604. We are overdue for another. The Rare Book and Manuscript Library, University of Illinois.

stellar positions, which led Kepler to devise his laws, which led to Newton, and finally to our own time. As Tycho lay dying in Prague some three decades later, Johannes Bayer was recording Tycho's work in his great *Uranometria*, within which he placed "Tycho's Star" even though it was long gone and by then a memorial to the great observer.

We now know that Tycho's Star was not "new," but old, a star in the act of violently dying, and not a "nova" as we call such things today, but a "supernova," an exploding star whose energy release dwarfs those of all other celestial events except variations of its own kind. Briefly and brilliantly lighting their immediate surroundings, supernovae can be detected telescopically from nearly the edges of the visible Universe, billions of light years away.

Chapter 4 began to discuss supernovae, Chapter 5 told which of the first magnitude stars — Antares, Betelgeuse, several others — could become one. Now it's time to see what they are really like, and what we might expect someday to see. Why though, was it necessary to change the name? Why ditch "nova" for "supernova"? Because there *are* such things as "ordinary" novae. While lesser creatures than supernovae, they can easily — and far more frequently — hit first magnitude and thereby change the visible sky. And some are precursors to whopping supernovae. So we hit them first.

Nova

Quick now quiz-show fans, except for being bright, what do Spica, Antares, Rigel, Procyon, Sirius, Regulus, Acrux, Hadar, and Alpha Centauri all have in common? Answer: They are all at least double. (We should count Castor too.) What then ties Sirius, Procyon, and Regulus together? A: They all have white dwarf companions. Admittedly, it's an unlikely scenario. But it gets the point across. A lot of stars are at least double, many triple, quadruple, quintuple, even sextuple. Maybe half the stars you see are double or multiple, the factor changing with mass and class.

The origins of stellar duplicity are still shrouded in some mystery. Doubles are clearly born from the same birth cloud. They might simply collapse together as an orbiting pair rather than as a single star. Or a star might have a thick surrounding disk and, rather than growing planets, the disk might accumulate into a second star. Even capture of one star by another within a dense cluster might take place. However, it happens, doubles are certainly very common.

Once formed, there are more scenarios than one can count, each leading to different endings as the companionable stars age. We can only sample them. If the members of a double star are far enough apart, each goes through its evolution independently of the other. In all cases, the more massive one begins to evolve first (as pointed out in Chapter 4). If both are above the 9 or so solar mass limit, then they both develop iron cores and explode, one after the other, leaving us with perhaps a double neutron star (observed!), or a neutron star-black hole combo (possibly observed), or even a pair of black holes (not observed, so far at least; how would one know?). It's also possible that if the stars are close enough that the first explosion will eject the companion, which goes off as a high-speed "runaway star." Such are also observed. Betelgeuse (Chapter 5) might be an example.

If they are both under the 9 solar mass limit (ignoring the intermediate case of one above, one below), then the more massive evolves first as a giant and turns into a white dwarf. Here we are back on the familiar ground trod by Sirius, Procyon, and Regulus. These bright stars, which currently dominate their systems, were once the lesser components. Now the first magnitude stars we see, each of these have white dwarf companions that were once the greater of the two, but are visible only with good optical aid.

Now change the initial conditions. Squeeze the original pair close together at birth. As the more massive member expands into a giant (either with a dead helium core or as a more evolved giant with a dead carbon core), it will encroach upon the other, and we have another variety of possible acts. In some cases, if the lesser star, still a dwarf, is massive enough, it can raise tides in the developing giant. Matter can then flow from the giant toward the dwarf, which accumulates mass at the expense of its evolving companion. This is what is happening to the star Algol (Beta Persei) in the constellation Perseus. The classic "eclipsing double," every 2.87 days a distended orange giant orbits in front of a massive blue dwarf for a few hours, cutting the light from the system from second magnitude to third. (Such eclipsers are actually quite common.) The barcode spectrum of the system is a combination of those of the two components. From the Doppler effect we can derive velocities and along with data from the eclipses, the complete orbit. From Newton's version of Kepler's laws, we then measure the masses to find that the evolved giant is contrarily *less* massive than the dwarf. But it *had* to be the more massive at one time. It's being robbed of its substance and perhaps will even be destroyed.

But what happens if the secondary dwarf is of much lower mass, with insufficient gravity to pull matter from its more massive live-in expanding neighbor? One possibility is that the growing giant, or its dense wind, will simply encompass the dwarf, which will cause the two to spiral slowly closer. If conditions are just right, when the more massive star has turned into a white dwarf, it's got a little, low mass, red dwarf, or even solar type star, squeezed in right next to it.

The situation becomes reversed. The dense dead white dwarf is now close enough to raise tides in the remaining ordinary *red* dwarf, and matter flows in the other direction, *on* to the evolved star rather than *from* it. Theory and observation show that mass does not flow directly, but into a disk around the white dwarf, from which it settles down onto the surface. White dwarfs, which stripped themselves of their outer envelopes when they were giants, have virtually no hydrogen left, and no fusion to speak of. But with the new closeness of the pair, along comes some more hydrogen, manna from heaven of a sort. The white dwarf slowly builds up a new hydrogen surface layer. The compression of the layer by the white dwarf's fierce gravity raises the temperature at the bottom, which goes higher and higher until it hits the hydrogen fusion point. The layer then flames out in a nuclear explosion. And a *nova* is born, the explosion increasing the double star's visual brightness by hundreds of thousands, even millions, of times, the star's luminosity becoming 100,000 or more times that of the Sun.

And Here They Are

Though far less powerful than supernovae, ordinary novae make up for it by their sheer numbers, meaning that a few will be nearby and quite bright. Every year has its set of faint or telescopic ones, and maybe there are 40 in our Galaxy annually, most of them hidden by dust or distance. But put one of these white dwarf surface-exploders close enough, make it sufficiently energetic, and it can hit the goal of first magnitude, making it eminently visible from just about anywhere, depending of course on where it is relative to the sky's equator. The twentieth century saw five such, oddly all in the first half, but if you don't mind stretching the definition to bright second (à la Castor), we can add another in the century's second half. That's an average of one every 17 years. Of the six, five were visible from the temperate north, giving us one locally every 20 years, which makes it not just possible, but probable, that you will get a chance to see one, or — given that averages don't mean much here — two or three. (Or for that matter, none at all.) The duration near maximum — the "hang time" — depends on the details of the parent double star. The more energetic "fast" novae fade by several magnitudes within a few days, requiring that you have clear skies and look immediately, while the slowest ones can take weeks or even months to settle down, letting you wait out clouds for a clear view. And there is everything in between.

The best known nova is probably Nova Herculis of 1934 (also known by its variable-star name, DQ Herculis), which in December of that year shone at magnitude 1.5 among the stars of the northern part of the constellation Hercules, nine degrees northwest of Vega. Though just barely making the first magnitude cut, it faded only slowly, allowing any interested party to watch for some time as it declined through second and third magnitude. Estimated to be over 1500 light years away, if placed at Vega (25 light years distant), it would have packed the light of 15 that of our Venus.

The Nova Herculis explosion was relatively weak. Near the other end of the true-luminosity scale, the blast from Nova Puppis (in the southern constellation, Puppis, the Keel of the Ship Argo) hit us early in November of 1942. Flickering briefly at magnitude zero, it rivalled Alpha Centauri in brightness as it made a giant triangle with Sirius to the northwest and Canopus to the southwest. Though 35 degrees south of the equator, thanks to the timing, the nova was still visible throughout much of the northern hemisphere just before sunrise: if anybody cared to look, what with World War II well underway. Near 1600 light years away, if Nova Pup had been as close as Vega, it would have not just overwhelmed Venus, but would have been as bright as a quarter Moon.

The brightest to the eye of all the twentieth century was Nova Aquilae of 1918, which lit the sky 17 degrees to the southwest of Altair. Hitting magnitude −1.1, roughly between the brightnesses of Sirius and Canopus, it shone half a dozen times brighter than Aquila's first magnitude luminary.

I've seen a few novae, studied one, took amateur photographs of another. But the one that stands out is Nova Cygni 1975. While it did not quite make our criterion of first magnitude, at bright second (1.8) it came close, close enough that perhaps as for Castor we might give it the title of "honorary first." It does, however, have real honor of its own, the title of the fastest nova on record. Just a few degrees north of Deneb, perfectly

Figure 7.2   Nova Aquilae 1999 (circled) lies to the right of the familiar three-star pattern of Aquila (the Eagle) as seen at lower left in Figure 5.4. While reaching just fourth magnitude and visible to the naked eye only from a dark location, it is typical of the occasional bright nova, one that can hit first magnitude or beyond, thus changing the whole pattern of it's constellation.

timed in August of 1974, for a brief astronomical moment the Swan appeared to have two tails. In absolute terms, it was one of the more luminous known.

Gradually, over the months and years, the brilliance of the nova disappears into the darkness of the sky. After 75 years, Nova Herculis sank 300,000 times fainter than its maximum brilliance, returning pretty much to its original status. After just a few decades, Nova Cygni was down to under a millionth. For any nova, several years after the initial outburst, the blasted surface layer, ejected at 1000 kilometers per second, becomes telescopically visible as an illuminated expanding shell. Many of them are enriched in chemical elements created in, and dredged up by, the nuclear detonation, which adds to the load of heavy stuff in the star-forming interstellar gases. Eventually, long after the white dwarf companion has returned to normal, the

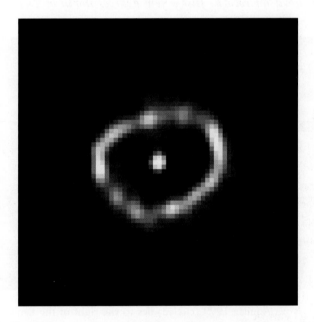

Figure 7.3   Expanding at a thousand kilometers per second, the exploded surface of Nova Cygni 1992 reached a diameter of 1000 AU, 25 times the size of Pluto's orbit, in only seven months. At the center, the highly disturbed white dwarf shines brightly. Hubble image, F. Paresce, R, Jedrzejewski (STScI), NASA/ESA.

whole accretion process starts all over again, only to be blasted off once more 100,000 or so years later (or so we might guess, the true number hardly known from observation).

While there are lots of stars with white dwarfs tucked in close to them, we know of no actual nearby candidates. Prediction is quite impossible. The white dwarfs that belong to Sirius and Procyon are far too distant from their first magnitude companions to be able to accrete much if any mass from them. But that means nothing. A first magnitude nova is always not just possible, but likely once a generation or so. With luck, we'll all get to see another one or two of them, maybe one that will bring the children out to look, one or more intrigued enough to become an astronomer.

Supernova

Now we have just to be plain lucky. While the chances of seeing a first magnitude (or brighter) nova in a lifetime are high, the odds of admiring a similarly bright supernova are not. If we include Nova Cygni 1975, the twentieth century saw 6 first magnitude novae. The historical record shows the same number of first magnitude supernovae not in a century, but in two millennia! Which reduces the odds that you will get to see one by a factor of 20, for an average of one every 300 or so years, four times a good human life span.

As we saw before, though, when it comes to small numbers of things, averages don't mean very much. These spectacular events can clump together and then we can go for a long time with no bright supernovae at all, which is what is happening now. The last two took place in 1572 (Tycho's Star) and 1604 (Kepler's), just 32 years apart. It was then 383 years until the next one, Supernova 1987a, and that was in a neighboring satellite galaxy (the Large Magellanic Cloud) that reached just third magnitude because of its huge distance of 170,000 light years. Which, by the way, shows just how powerful these things can be!

Four galaxies (not counting ours) are visible to the naked eye. In the north is the prominent (at least in a dark sky) fourth magnitude

Andromeda Galaxy (in the constellation of the same name). The farthest thing the human eye can easily see, its light left there more than two and a half million years ago. It's comparable to our own Galaxy, and in 1885 hosted its own supernova (now known as S Andromedae). Just at the edge of human vision, were it not for the background brightness of the galaxy itself, the supernova might well have been visible without a telescope. Not far away across the sky, in the more obscure constellation Triangulum, lies a smaller, fainter, somewhat more distant galaxy that is a real challenge to view without optical aid. A pair in the deep southern hemisphere, the Magellanic Clouds, are small satellites of our Milky Way Galaxy, but you really need to be south of the equator to see them decently. Averaging 180,000 light years away, they orbit our own system and are slowly being torn apart by tides raised by the combined gravity of the stars in the Milky Way. The huge tidal disturbances promote star formation that includes gigantic clusters of massive stars, all of which have the potential to create supernovae. In the Large Cloud we in fact see generations of them, the blast from each one helping to compress the interstellar gases, which enhances star formation, which leads to more supernovae, and on and on. We see the same thing happening in our own Galaxy, the others teaching us about ourselves.

The remaining four known historical supernovae, those that pre-date Tycho's star, are — like ancient comets — known to us mostly from old Chinese commentaries by scholars and astrologers who assiduously recorded celestial events. They stand out from other phenomena for their extreme brightness, which can put that of ordinary novae to shame. Tycho's Star hit magnitude −4, six times brighter than Sirius, while Kepler's reached −3, still well outside the normal stellar realm. We have to go back another 390 years to get to the next one in the past, the Supernova of 1181, which oddly also showed up in Cassiopeia just 10 or so degrees to the east of where Tycho's would be seen four centuries later. The old record suggests a magnitude of −3, comparable to Kepler's Star (and Jupiter) but not quite reaching Tycho's.

Then we go back just 127 years to the one that is, after Tycho's perhaps, the most famed of all supernovae, the "Chinese Guest Star" of the year 1054. Appearing just a degree to the northwest of Zeta Tauri, the star that marks Taurus the Bull's southern horn, 15 degrees east-northeast of first magnitude Aldebaran, the Guest Star also hit magnitude −4. Again comparable to Venus, it too was visible in full daylight and took a year to fade away. There are various recordings of it around the world, one possibly a stone pictograph made by the natives of the American southwest. But there is nothing significant in Europe, where it was surely seen. However, can you imagine someone not only being out at night in the 11th century, but saying in public that the heavens had changed? "Maybe if we don't say anything, it'll go away." Which it did.

Then another two in a row. Just 48 years earlier, in 1006, was the whopper of the past 2000 years, the brightest for which we have written records. Far south (42 degrees below the equator) and tough to see from the north, it fell right on the modern boundary between the constellations Lupus (the Wolf) and Centaurus some two dozen degrees southwest of Antares. From the descriptions, the brightness is hard to reconstruct. One scribe wrote that "things could be seen by its light," and the most likely estimate gets it as bright as magnitude −8, which is about that of a mid-crescent Moon, though some authorities suggest it was even brighter.

Finally, we jump another long gap into the past, more than eight centuries, from 1006 to the year 185, whose supernova was roughly again at −2. This one, appearing just a couple degrees south of the Alpha Centauri, would have been visible only from far southern China. It took place so long ago that records are unclear. Given the haze of the distant past, other supernovae may have surfaced within that huge time span. With an average of one every third of a millennium, we are just barely overdue for another. Which again means nothing at all.

Though all of these supernovae are thousands of light years away, they have the power to cast shadows on Earth. A typical supernova can be billions of times more luminous than the Sun, nearly as bright as all the stars in its parent galaxy put together. Place a typical one at Vega's distance, and it would appear at magnitude −18 or −19, the latter 250 times the brightness of the full Moon. Such closeness is inside the so-called "death zone," within which a supernova would damage the Earth by stripping at least part of the atmospheric ozone layer. Dangerous energetic ultraviolet sunlight, which is blocked by ozone, could then cascade down to the ground, killing off part of the food chain and possibly initiating a mass extinction. It's probably happened, in fact there is some evidence that it *has* happened. But relax, as the closest candidates are Betelgeuse and Antares, and they are each around 550 light years distant. If either went off, each would very roughly give us a benign one gibbous Moon's-worth of light.

Supernovae are rare in our Galaxy, but not in the Universe at large. These explosions are so bright that they can be seen billions of light years away. There are so many galaxies, vast numbers of them, within a sphere of that size that supernovae are really quite common. Just pick a number. Say ten million or more galaxies are close enough for study. If each one kicks off even just one per century, that gives us 100,000 supernovae per year, or roughly 300 supernovae per DAY, one every five minutes. We are not yet able to patrol the whole sky on a nightly basis, but dedicated wide-angle searches show supernovae to be numerous indeed. If everything were visible, the Universe would appear to sparkle with firecrackers as they popped off, one after the other, a deeper look showing the nova-like fireflies that are a thousand times more frequent.

The Crab Walks at Midnight

The historical records are difficult to follow and descriptions are approximate at best, if not downright inadequate. How then do we know exactly where in the sky the supernova went off? For that we turn to the 18th century when astronomers were beginning to cata-logue the sky. The greatest of them was William Herschel, the trans-planted (from Germany) English astronomer who in 1781 discovered the planet Uranus, and who with his sister Caroline found immense numbers of interstellar nebulae (bright dust and gas clouds illumi-nated by nearby stars), clusters, doubles, and more.

But there were others, among them a contemporary of Herschel, Charles Messier of France. Known as a comet hunter (with 13 to his

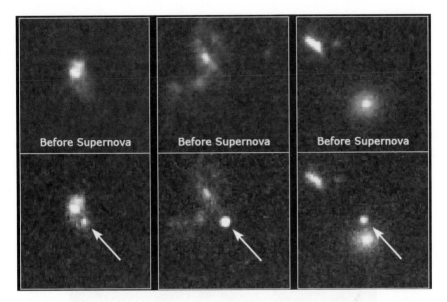

Figure 7.4 The power of a supernova rivals that of its parent galaxy as seen in this trio of ultradistant systems. Hubble images, NASA and A. Riess (STScI).

credit), he compiled a list of 103 fixed celestial objects that through the telescope might fool an observer into thinking that he or she (Caroline Herschel had a number to her credit as well) had found a comet. The Messier list includes some of the best known and beloved of heavenly sights, including (probably for completeness) the Pleiades star cluster as Messier 45, Cancer's Beehive cluster as M 44, as well as fainter clusters, nebulae, and external galaxies (whose natures were not at the time known); for the most part anything "fuzzy" and not a star. Other astronomers appended a few more objects, bringing the modern Messier list to 110.

Messier's Number 1 is a strange looking gaseous nebula easily visible in a small telescope. A tenth of a degree across, it appears as a uniform glow interlaced with crab-like tentacles, from which it got its popular name, the "Crab Nebula." It's right off Taurus's southern horn, right where the Chinese drew their famed Guest Star of 1054. Over the years it's been observed to be growing larger, and is now some ten light years across. Tracing the expansion back into the past

Figure 7.5   Among the favorites of the sky is the Crab Nebula, the 10-light-year-wide expanding remains of the great supernova of the year 1054. Blowing up in western Taurus, at its peak it was visible in full daylight. At dead center lies a neutron star roughly the size of Manhattan (see Chapter 4). Spinning 30 times per second, carrying perhaps double the mass of the Sun, it powers the entire affair. The supernova remains are sending newly-created elements back into the cosmos from which it came. NASA, ESA, J. Hester and A. Loll (Arizona State U.).

gives a starting date in the 11th century. Currently expanding at 1500 kilometers per second (three million miles per hour!), the Crab is without question the exploded debris of the supernova.

At its center is a faint star that turns itself on and off 30 times per second. Astronomers are convinced that it is a spinning neutron star, the collapsed remains of a once-magnificent high mass star that had probably first turned itself into a red supergiant, perhaps one like Betelgeuse or Antares, though from Earth the supernova's progenitor would have appeared fainter than the naked eye limit. Then its nuclear-burning core, which had finally turned to iron, collapsed, and the star shot up in brightness by tens of thousands of times to become over a billion times more luminous than the Sun, so bright that even though 6000 light years away, it was visible in the daytime. Now a "pulsar," the little star beams intense radiation along a magnetic axis that is tilted against the rotation axis. Every time the wobbling magnetic pole points toward us we get a burst.

The strength of the explosions and their leavings goes far beyond simple bright lights in the sky. Supernovae produce vast shock waves in the interstellar gases that are so powerful they accelerate atomic nuclei (atoms stripped of their electrons) to nearly the speed of light, whence they become what we call "cosmic rays." Circulating through the Galaxy, trapped by its magnetic field, a tiny fraction hammer into Earth's atmosphere. The colliding atomic debris, which at maximum can carry the energy of a pitched baseball, produces the heavy form of carbon (C-14) used in archeological radioactive dating and may even in part be responsible for triggering cloud formation and lightning.

All the historic supernovae have left such gaseous remains. Some are only barely recognizable as such, but they are all there, and are especially visible from their production of radio waves. For years the reality of the supernova of 185 was uncertain, and has only recently been confirmed, in part through observation of X-ray radiation from the expanding shell. Most of these gaseous remnants are not the debris of the actual explosion so much as the lingering expanding shock waves in the interstellar gases that radiate across the spectrum. In addition, there are hosts of others from supernovae that were too far away to be seen with the naked eye, or hidden behind dark dust clouds that their radio waves can penetrate. When we take these into account, our Galaxy produces two or so supernovae per century. But of two remarkably different kinds.

Double Down

All the massive stars of Chapter 5, those above 9 or so solar masses (which are marked with asterisks in the Quick Locator of Chapter 5), have the potential to develop iron cores and to explode as supernovae like the Chinese Guest Star of 1054 and Supernova 1987a, which ripped off not all that long ago in the Large Magellanic Cloud. There is, however, an equal route to glory, one first encountered in Chapter 4.

"Ordinary" novae exhibit a great variety of qualities that depend on the masses of their orbiting stars, as well as on their orbital separations and the states of the companions that donate mass to their captive white dwarfs. The vast majority of white dwarfs, whether in double systems or not, have masses well beneath that of the Sun. A small handful of close pairs, however, will have white dwarfs close to the allowed limit of 1.4 solar masses. When the fresh hydrogen layer that has been deposited on the white dwarf gets blown off in a nova explosion, a bit of the donated matter is thought to be left behind, causing the white dwarf's mass to increase. Repeated detonations cause a steady rise until the white dwarf reaches the maximum value and can no longer support itself. When it begins to collapse, it heats, and then violently flames out as the whole thing, all 1.4 solar masses, explodes in a gigantic "carbon bomb." Nothing, no stellar remnant, is left at all, though the mass-donating companion may well survive. Good candidates are several recurrent novae that go off every few decades, have high mass white dwarfs, and evolved donors that yield a good supply of fresh hydrogen (though the relation is contended).

There is another good possibility in which the components of the double star are *both* white dwarfs. When each was ageing as a giant, it encompassed the other, which brought the final products very close together indeed. If close enough, they could radiate away their orbital energy as gravity waves in Einstein's spacetime (Chapter 3). Eventually, they merge. If the sum of their individual masses is greater than the white dwarf limit, off it goes. Both mechanisms may be responsible.

No matter the details of their origins, white dwarf supernovae are typically two to four times more powerful than the standard massive-star core-collapse version. We can tell the difference between the two by their barcode spectra in that the core collapse

version has a signature of hydrogen from the extensive outer envelope that gets blasted into space, whereas the white dwarf variety
does not. Moreover, since massive stars do not live long, they and
their supernovae are confined to the galaxies' disks where stars are
now being born. The white dwarf variety, however, can also be
found in the old outer halos of galaxies where few massive stars
tread, as well as in galaxies that have no disks or active star formation. They can be seen for billions of light years away and, because
of their uniformity, are profoundly important in establishing the
nature of the expansion of the Universe. The white dwarf version is
not only typically brighter than the core-collapse kind, but also produces three times as much iron, about a third of a solar mass.
Without them, without double stars and white dwarfs (as well as the
core-collapsers, which seem to be the more numerous), there may
never have been enough iron and other metals in the Universe to
have created the Earth.

Close to home, it's often hard to tell which kind produced the
historical supernovae, as all we often have is a brightness estimate and
the current nature of the expanding gaseous remnant. The big exception is the Guest Star of 1054, for which we have the leftover neutron
star as supreme evidence. And in modern times, we know Supernova
1987a in the Large Magellanic Cloud was caused by massive-star collapse as the progenitor supergiant was quickly identified on earlier
photos taken of the region. (Its neutron star, if there is one, has never
been found.) Kepler's and Tycho's are believed to have been caused
by white dwarf explosions (Kepler's contended), as were the events of
1006 and 185. The supernova of 1181, however, seems to have been
produced by massive-star destruction.

So at the end, the odds are not large that any of us will get to see
a supernova. We could go another 400 years. On the other hand, one
could as easily erupt tonight. Statistics being what they are, we could
even have two within just a few years. Whatever the time scale, a first
magnitude supernova is still inevitable. When we do get one, you
won't have to look for it. It will be right there in front of you, even
from the streets of the Big City, and the media will without question
let everyone know about it.

Eruption

To the eye, Eta Carinae (which shares its constellation Carina with first magnitude Canopus) is an innocuous fifth magnitude star buried within a spectacular region of the southern Milky Way that is filled in with bright and dark interstellar clouds and star clusters. Not so in the 1840s, when the star underwent "the biggest non-terminal eruption we know about" [K. Davidson]. Novae and supernovae aside, Eta Car is the Mount Aetna of stars. "Normally," if that is a decent word for a star like this, it probably hovers around fourth magnitude, not much brighter than the mid-handle stars of the Little Dipper. From the sketchy information we have, it brightened some in the 1700s, but then around 1830 it really took off, reaching and sustaining first magnitude for around 25 years between 1835 and 1860. In 1842 it achieved −1, falling in brightness between Sirius and Canopus, which briefly made it the second brightest star in the sky. Unfortunately, it's 60 degrees south of the equator, and cannot be seen by anyone much above the Tropic of Cancer. But what a sight for southerners, with the three brightest stars of the sky making a crude, giant triangle.

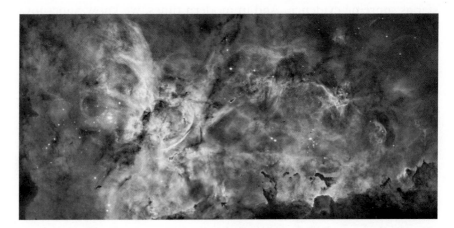

Figure 7.6   The spectacular Carina Nebula, some three degrees across and visible to the naked eye from deep southern locations, is filled with young clusters and massive new stars. Eta Carinae (Figure 7.7) is the brightest body to the left of center. NASA, ESA, N. Smith (U. Cal. Berkeley), and the Hubble Heritage Team (STScI/AURA).

Ejecting a huge, dusty, near-opaque gas cloud of perhaps a solar mass, Eta Car just pulled its own blanket over its head and disappeared. By 1865 it had fallen to mid-sixth magnitude, and it did not recover to naked-eye view until the latter part of the twentieth century. In the meantime, the erupted cloud has swelled to nearly a light year across. The energetics of the event are unique, though similar lesser action has been seen for a handful of other stars. We have no idea why or how the eruption took place, or how often it happens for any particular star.

We do know, however, that Eta Car is one of the most massive stars in the Galaxy. Though lighter now because of its eruptions and powerful mass-losing wind, it's thought to have been born near the upper range of stellar possibility, around 100 solar masses or even more. It also has a massive companion that orbits every 5.7 years, and how that's related we don't know either. Like supernovae, we can't predict when a massive erupter will go off and perhaps reach first magnitude. But it's one more possibility we can add to novae and supernovae. Not to mention adding all these to the far more numerous fireball meteors and comets of the last chapter.

Rare As They Come But Oh What a Sight One Would Be

Eta Carinae leads us into the realm of the most massive of stars. As we climb the mass ladder, stars get amazingly rarer, only one in a million being above the mass limit that makes it capable of becoming a supernova. At the top, above 100 solar masses, perhaps as much as 200, there might be only a handful of stars in an entire galaxy. Now we also enter the arena of great speculation. The more common stars in the upper range of our "first magnituders" of Chapter 5, those like Betelgeuse, Antares, Acrux, most likely will collapse to neutron stars. But these tiny bodies have limits too that seem to hover around three times the mass of the Sun. Astronomers think that rare stars above a high limit, perhaps 40 or 50 solar masses (but one really unknown), collapse to remnants that are too massive to be neutron stars, and that these in turn continue the collapse into black holes. The supernovae

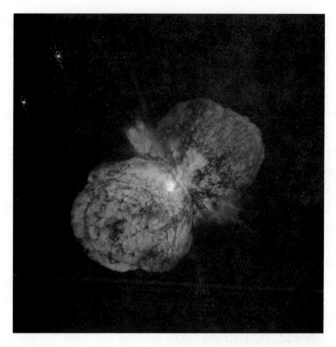

Figure 7.7. Eta Carinae erupted in the mid-nineteenth century to become one of the brightest stars in the sky even though six-eight thousand light years away. Barely visible, it is now buried within its own ejecta. Its fate may be to explode and produce a gamma ray burst visible for billions of light years; see Figure 7.8. Hubble image: Jon Morse (U. Colorado) and NASA.

that make them may be spectacular, brighter than anything we have ever seen so far in our Galaxy.

Speculation is one thing, observation another. And it's there. In 1967, a defense satellite designed to look downward for the gamma ray signatures of atomic bomb explosions instead saw one on high, from the sky. By the end of the century, dedicated astronomical satellites were seeing one of these intense bursts a *day*. As usual, there are two different kinds, which divided at two seconds duration. Eventually, telescopes on the ground were able to relate the longer version of these gamma-ray bursts to optically bright "afterglows" that, when they cleared, revealed ultra-intense supernovae at their cores: *hypernovae* that can be more luminous than even white dwarf detonations. They may be caused by supermassive stars like Eta Carinae that are collapsing into black holes. Even at distance of six to eight thousand light years, if Eta Carinae were to go off, the gamma ray blast could be dangerous to Earth. The so-far observed champion was one that was visible to the naked eye even though it was estimated to be 7 billion light years away.

We are saved largely by the structure of these hypermassive explosions. The only way to reconcile the energetics of gamma ray bursts is to theorize that the radiation is beamed outward in cones that center on the stars' rotation axes. From the shape of the dusty cloud around Eta Car, the rotation axis is pointed in a direction not at us. It's somebody else's problem. We have no idea of how often we might get hit. Scientists have speculated that at least one was the cause of a mass extinction. A crude estimate is once in a billion years, which gives us five such events. The estimate also includes never.

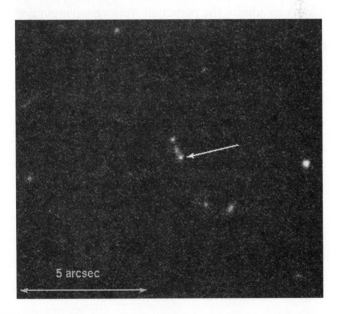

Figure 7.8   Looking like an ordinary star, it's really the optical afterglow caused by a gamma ray burst billions of light years away. Such glows can be bright enough to be seen with the naked eye. One close enough could overwhelm the sky. Hubble image, D. W. Fox, J. S. Bloom, S. R Kulkarni (Caltech), *et al.*

Then there are the short bursts to deal with, which are more or less guessed to be mergers by double neutron stars or double orbiting black holes, or a combo of the two in double systems. A close one of these would not do us any good either.

"First Magnitude"

Gives a whole different perspective to the sky. Even in a light-polluted environment, there is so much to see, from the motion and effects of the Sun and Moon, through the Moon's changing shape, to the wanderings of the bright planets, which leads us inevitably to the stars, to the ones of constant brightness, then to the exploders and erupters.

Once seen, all these bright celestial sights guide us into the darkness on a quest to look ever deeper into the cosmos, to where you can see the Milky Way, even other galaxies, as we pause in our evening activities to go outdoors to admire the perpetual, yet ever-changing, sky.

INDEX